面向高等职业院校基于工作过程项目式系列教材
企业级卓越人才培养解决方案规划教材

基于 Vue 的 Java Web 项目实战

天津滨海迅腾科技集团有限公司　编著

图书在版编目(CIP)数据

基于Vue的Java Web项目实战 / 天津滨海迅腾科技集团有限公司编著. —天津：天津大学出版社，2020.9（2022.1重印）

面向高等职业院校基于工作过程项目式系列教材. 企业级卓越人才培养解决方案规划教材

ISBN 978-7-5618-6771-6

Ⅰ.①基… Ⅱ.①天… Ⅲ.①JAVA语言－程序设计－高等职业教育－教材 Ⅳ.①TP312.8

中国版本图书馆CIP数据核字(2020)第176761号

JIYU Vue DE Java Web XIANGMU SHIZHAN

主　编：齐立辉　周仲文
副主编：闫瑞雪　王晓英　张明宇
　　　　李世强　翟亚峰

出版发行 天津大学出版社
地　　址 天津市卫津路92号天津大学内(邮编:300072)
电　　话 发行部:022-27403647
网　　址 www.tjupress.com.cn
印　　刷 廊坊市海涛印刷有限公司
经　　销 全国各地新华书店
开　　本 185mm×260mm
印　　张 17
字　　数 462千
版　　次 2020年9月第1版
印　　次 2022年1月第2次
定　　价 69.00元

面向高等职业院校基于工作过程项目式系列教材
企业级卓越人才培养解决方案规划教材
编写委员会

杨　峰　山东胜利职业学院
成永江　东营科技职业学院
刘文娟　德州职业技术学院
杜卫东　枣庄职业学院
常中华　青岛职业技术学院
刘　磊　临沂职业学院
董红兵　威海海洋职业学院
李秀敏　烟台汽车工程职业学院
宋　军　山西工程职业学院
刘月红　晋中职业技术学院
田祥宇　山西金融职业学院
赵　娟　山西旅游职业学院
陈　炯　山西职业技术学院
范文涵　山西财贸职业技术学院
李艳坡　河北对外经贸职业学院
杨海源　衡水职业技术学院
娄志刚　唐山科技职业技术学院
刘少坤　河北工业职业技术学院
尹立云　宣化科技职业学院
孟敏杰　许昌职业技术学院
李庶泉　周口职业技术学院
周　勇　四川华新现代职业学院
周仲文　四川广播电视大学
邱　林　天府新区通用航空职业学院
贺国旗　陕西工商职业学院
夏东盛　陕西工业职业技术学院
景海萍　陕西财经职业技术学院
许国强　湖南有色金属职业技术学院
许　磊　重庆电子工程职业学院
谭维齐　安庆职业技术学院
董新民　安徽国际商务职业学院
孙　刚　南京信息职业技术学院
李洪德　青海柴达木职业技术学院

基于产教融合校企共建产业学院创新体系简介

基于产教融合校企共建产业学院创新体系是天津滨海迅腾科技集团有限公司联合国内几十所高校，结合数十个行业协会及1000余家行业领军企业的人才需求标准，在高校中实施十年而形成的一项科技成果，该成果于2019年1月在天津市高新技术成果转化中心组织的科学技术成果鉴定中被鉴定为国内领先水平。该成果是贯彻落实《国务院关于印发国家职业教育改革实施方案的通知》（国发〔2019〕4号）的深度实践，开发出了具有自主知识产权的“标准化产品体系”（含329项具有知识产权的实施产品）。从产业、项目到专业、课程形成了系统化的操作实施标准，构建了具有企业特色的产教融合校企合作运营标准“十个共”，实施标准“九个基于”，创新标准“七个融合”等全系列、可操作、可复制的产教融合系列标准，取得了高等职业院校校企深度合作的系统性成果。该成果通过企业级卓越人才培养解决方案（以下简称解决方案）具体实施。

该解决方案是面向我国职业教育量身定制的应用型技术技能人才培养解决方案，是以教育部—滨海迅腾科技集团产学合作协同育人项目为依托，依靠集团的研发实力，通过联合国内职业教育领域相关的政策研究机构、行业、企业、职业院校共同研究与实践获得的方案。该解决方案坚持“创新校企融合协同育人，推进校企合作模式改革”的宗旨，消化吸收德国“双元制”应用型人才培养模式，深入践行基于工作过程“项目化”及“系统化”的教学方法，形成工程实践创新培养的企业化培养解决方案，在服务国家战略——京津冀教育协同发展、“中国制造2025”（工业信息化）等领域培养不同层次的技术技能型人才，为推进我国实现教育现代化发挥了积极作用。

该解决方案由初、中、高三个培养阶段构成，包含技术技能培养体系（人才培养方案、专业教程、课程标准、标准课程包、企业项目包、考评体系、认证体系、社会服务及师资培训）、教学管理体系、就业管理体系、创新创业体系等，采用校企融合、产学融合、师资融合“三融合”的模式在高校内共建大数据（AI）学院、互联网学院、软件学院、电子商务学院、设计学院、智慧物流学院、智能制造学院等，并以“卓越工程师培养计划”项目的形式推行，将企业人才需求标准、工作流程、研发规范、考评体系、企业管理体系引进课堂，充分发挥校企双方的优势，推动校企、校际合作，促进区域优质资源共建共享，实现卓越人才培养目标，达到企业人才招录的标准。该解决方案已在全国几十所高校实施，目前形成了企业、高校、学生三方共赢的格局。

天津滨海迅腾科技集团有限公司创建于2004年，是以IT产业为主导的高科技企业集团。集团业务范围覆盖信息化集成、软件研发、职业教育、电子商务、互联网服务、生物科技、健康产业、日化产业等。集团以科技产业为背景，与高校共同开展“三融合”的校企合作混合所有制项目。多年来，集团打造了以博士研究生、硕士研究生、企业一线工程师为主导的科研及教学团队，培养了大批互联网行业应用型技术人才。集团先后荣获全国模范和谐企业、国家级高新技术企业、天津市“五一”劳动奖状先进集体、天津市“AAA”级劳动关系和谐企业、天津市“文明单位”、天津市“工人先锋号”、天津市“青年文明号”、天津市“功勋企业”、天津市“科技小巨人企业”、天津市“高科技型领军企业”等近百项荣誉。集团将以“中国梦，腾之梦”为指导思想，深化产教融合，坚持围绕产业需求，坚持利用科技创新推动生产，坚持激发职业教育发展活力，形成“产业 + 科技 + 教育”生态，为我国职业教育深化产教融合、校企合作的创新发展作出更大贡献。

前　言

随着互联网的不断发展，Java Web 项目的需求不断扩大。本书主要以网上书城项目贯穿全书，对其涉及的相关知识进行讲解，包含其各模块的功能和使用方法，并基于 MVC 开发理念和 Vue 前端技术将 JSP 知识应用到实际的项目中。全书知识点由易到难，以“理论 + 实践”的形式展现给读者，学习完本书后，读者可以初步了解 Java Web 的理念和思想，并具备创建类似项目的能力。

本书主要内容包括 8 章，即网上书城项目环境搭建、登录页面设计、主页设计、业务对象封装、数据库链接、登录注册功能、应用 MVC 设计模式和 Vue 技术重构网上书城项目，严格按照生产环境中的操作流程对知识体系进行编排，循序渐进地讲解知识，并逐步应用所学知识。

本书中每章都设有学习目标、学习路径、任务描述、任务技能、任务实施、任务总结、英语角和任务习题，结构条理清晰、内容详细，任务实施可以将所学的理论知识充分地应用到实际操作中，提高读者的实践能力。

本书由齐立辉、周仲文担任主编，闫瑞雪、王晓英、张明宇、李世强、翟亚峰担任副主编，齐立辉和周仲文负责整书编排，第一章由齐立辉负责编写，第二章由周仲文负责编写，第三章由闫瑞雪负责编写，第四章由王晓英负责编写，第五章由张明宇负责编写，第六章由李世强负责编写，第七章由翟亚峰负责编写，第八章由张明宇和李世强负责编写。

本书理论内容简明扼要，实例操作讲解细致、步骤清晰，实现了理论与实践相结合，操作步骤后有相对应的效果图，读者可以直观、清晰地看到操作效果，牢记书中的操作步骤，能够更加顺利地学习 Java Web 相关知识。

天津滨海迅腾科技集团有限公司
技术研发部
2020 年 8 月

前　言

目 录

第一章　网上书城项目环境搭建

学习目标

通过学习网络交互模式，了解静态页面与动态页面的相关知识，学习 JDK 的安装与配置，掌握基本的 Web 知识，掌握开发环境的搭建，具有创建 Web 项目的能力。在任务实现过程中：

- 掌握 C/S 模式和 B/S 模式；
- 掌握静态页面和动态页面的相关知识；
- 掌握基本的 Web 知识；
- 掌握网上书城项目开发环境搭建。

学习路径

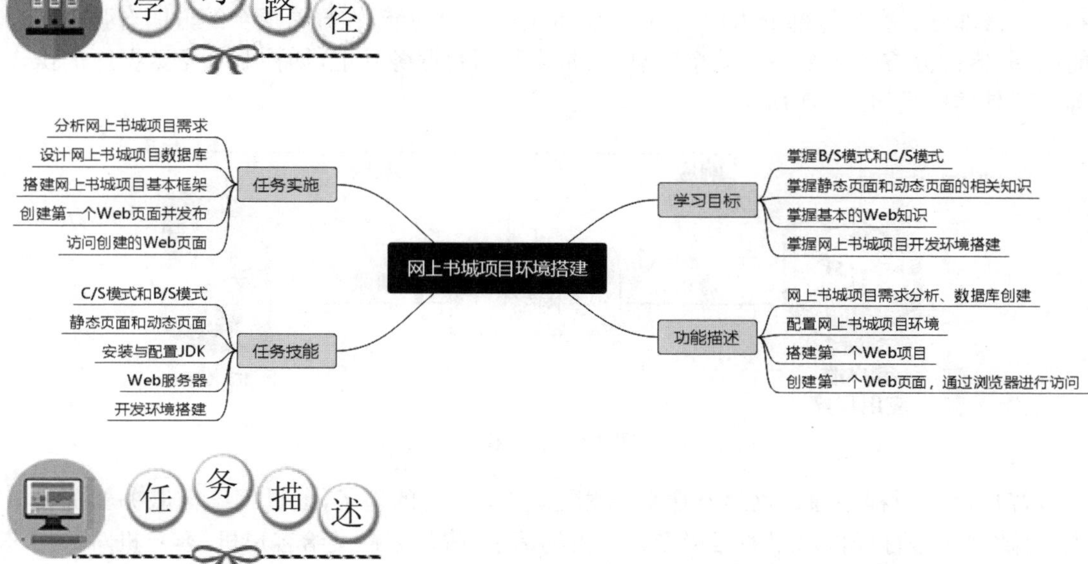

任务描述

【情境导入】

早期 Web 应用多为静态页面，不会根据用户的请求作出相应的回应。随着网络的不断发展，动态页面进入网络生活。纯静态页面已经不能满足大部分用户的需求，能时刻响应用户请求的动态页面已经成为主流。本次任务主要是新建 Web 应用，构建第一个动态 JSP 页面。

【功能描述】

- 网上书城项目需求分析、数据库创建；
- 配置网上书城项目环境；
- 搭建第一个 Web 项目；
- 创建第一个 Web 页面，并通过浏览器进行访问。

技能点一　C/S 模式和 B/S 模式

1.C/S 模式

客户机 / 服务器模式，也称 Client/Server（C/S）模式，由两部分组成，分别是客户机部分和服务器部分。客户机部分为用户专有，负责实现前台功能，在异常提示等方面有强大的功能；服务器部分方便多个用户共享信息，负责实现后台服务功能，如控制共享数据库的操作等。其具体模式如图 1-1 所示。

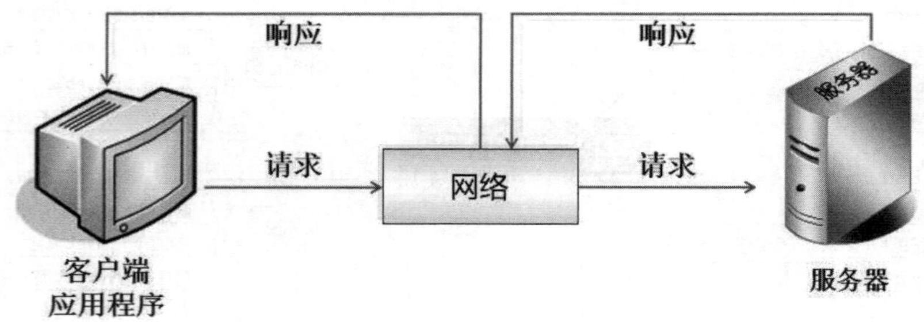

图 1-1　C/S 模式

客户机应用程序通过网络与服务器相连，接受用户的请求，通过网络向服务器提出请求，对数据库进行操作。服务器接受客户机的请求，将数据提交给客户机，客户机对数据进行计算并将结果呈现给用户。服务器还要对数据提供完善、安全的保护，保障数据的完整性，并允许多个客户机同时访问，这就对服务器硬件处理数据的能力提出了很高的要求。

C/S 模式主要用于客户机和服务器之间进行通信交流，具有以下优点。

- 交互性强。

- 存取模式安全。
- 响应速度快。
- 利于处理大量数据。
- 运行数据负荷较轻，数据的储存管理功能较为透明。

但是该模式也存在一定缺点，包括通用性低，系统维护、升级需要重新设计和开发，维护和管理的难度高，进一步拓展数据的困难较多，所以 C/S 模式只适用于小型的局域网。

2. B/S 模式

浏览器 / 服务器模式，也称 Browser/Server（B/S）模式，是 Web 兴起后的一种网络结构模式，用户主要使用的软件为 Web 浏览器。B/S 模式将系统的核心机制放置在服务器上，减轻了客户端的压力，使得系统在服务器上能够得到更好地维护、更新。客户端上只需一个浏览器即可通过 Web 服务器同数据库服务器进行数据交互。其具体模式如图 1-2 所示。

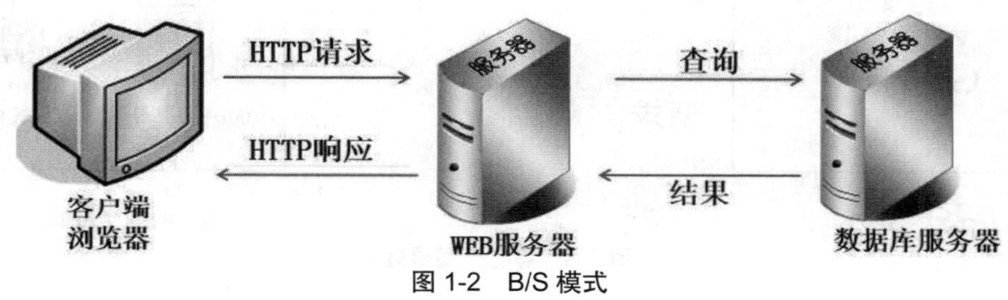

图 1-2　B/S 模式

B/S 模式最大的优点是总体拥有成本低、维护方便、分布性强、开发简单，可以不安装任何专门的软件就能实现在任何地方进行操作，客户端无须维护，扩展系统非常容易，只要有一台能上网的电脑就可以了。

C/S 模式和 B/S 模式都有自己的特色，但在硬件环境、安全要求、程序架构等方面存在不同，具体区别如表 1-1 所示。

表 1-1　C/S 模式与 B/S 模式的区别

	C/S 模式	B/S 模式
硬件环境	专用网络	广域网
安全要求	面向相对固定的用户，对安全的控制能力较强	面向不可知的用户群，对安全的控制能力较弱
程序架构	更加注重流程，对系统运行速度考虑较少	对安全以及访问速度都要多重考虑，是发展趋势
软件重用	差	好
系统维护	升级、维护困难	维护开销小，升级简单
处理问题	集中	分散
用户接口	与操作系统关系密切	跨平台，与浏览器相关
信息流	交互性强	交互密集

技能点二 静态页面和动态页面

1. 静态页面

Web 应用在早期主要是静态页面，页面主要使用 HTML 语言来编写，用户使用浏览器通过网络协议请求服务器上的 Web 页面，用户的请求被发送至服务器后经过服务器响应，再发送到客户端浏览器，显示给用户，如图 1-3 所示。

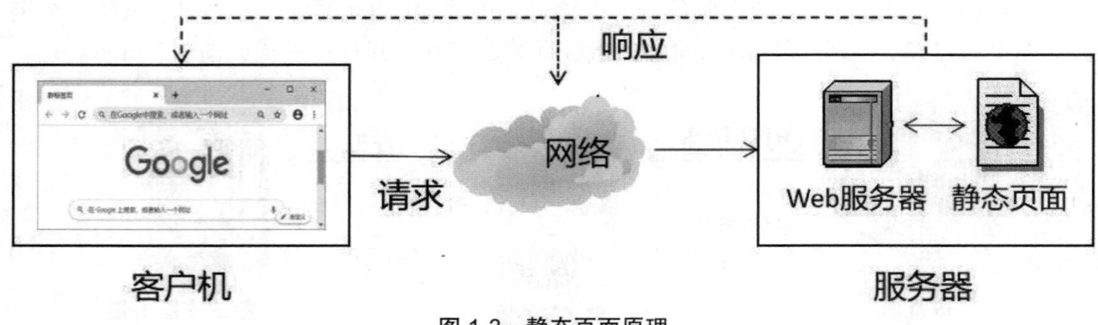

图 1-3 静态页面原理

2. 动态页面

Web 应用随着时代的发展变得复杂多样，相比早期 Web 应用大多是使用 HTML 语言的静态页面，现在更多的内容需要根据用户的请求动态生成页面信息，即动态网站。动态页面是由静态页面和动态脚本程序构成的，再将编写好的程序部署到 Web 服务器上，由服务器对程序进行编译，转化为浏览器可以解析的代码，返回给客户端浏览器，显示给用户，如图 1-4 所示。

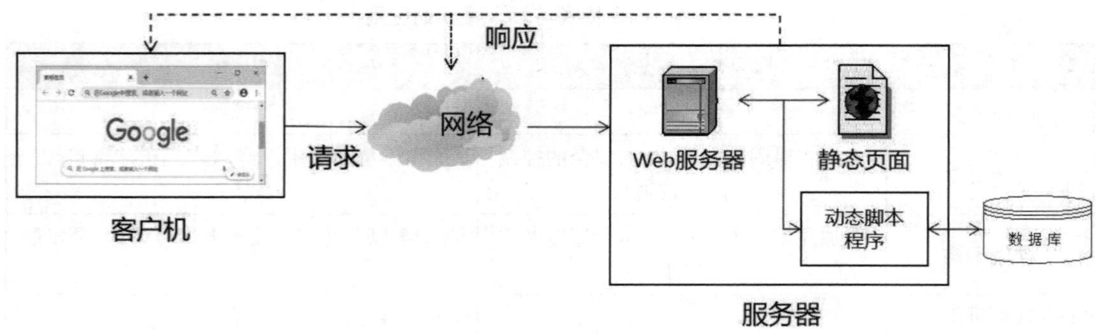

图 1-4 动态页面原理

3. 脚本语言

脚本语言是为了缩短传统编程语言的编程过程而创建的编程语言。在许多方面，脚本语言与高级编程语言互相结合交叉，没有明确的界限。

（1）ASP

ASP（Active Server Pages，动态服务器页面）是微软公司开发的服务器脚本环境，使用 VBScript 和 JScript 两种脚本语言，可以创建动态网页的 Web 应用。它没有自己专门的编程语言，允许用户使用已有的脚本语言编写 ASP 应用程序。但它提供了内置对象和 ActiveX 组件，利用这些可以拓展 ASP 业务功能。拥有这些能力，使得 ASP 的程序编写比 HTML 更加灵活、方便。ASP 图示如图 1-5 所示。

图 1-5　ASP 图示

ASP 的特点有以下几点。

● 提供了一些内置对象，增加了更多的功能。

● 可以运用 ActiveX 组件执行各种任务。

● ASP 在 Web 端运行，再将运行结果以 HTML 格式传送至客户端浏览器，使用者不会看到 ASP 所编写的原始代码，因此安全性更强。

（2）PHP

PHP（PHP：Hypertext Preprocessor，超文本预处理器）是一种在服务器端执行的嵌入 HTML 文档的脚本语言。PHP 有非常强大的功能，它能实现所有 JavaScript 的功能，几乎支持所有流行的数据库以及操作系统。PHP 图示如图 1-6 所示。

图 1-6　PHP 图示

PHP 的特点有以下几点。

● 开源免费。

● 拓展性强。

● 语言风格类似 C 语言，并且加入了面向对象的概念，简单的语法规则使它的操作编辑非常简单，实用性很强。

（3）JSP

JSP（Java Server Pages，Java 服务器页面）是由 Sun 公司于 1999 年 6 月推出的技术。它是在 HTML 文件中插入 Java 程序段（JavaScript）从而形成 JSP 文件。JSP 将网页逻辑与网页设计和显示分离，支持可重复使用的基于组件的设计，使基于 Web 的应用程序开发变得快速和便捷。JSP 图示如图 1-7 所示。

图 1-7 JSP 图示

JSP 的特点有以下几点。

- 继承了 Java 语言的相对易用性。
- 可跨平台使用。
- 可利用 JavaBean 和标签库技术重复使用常用的功能代码。

技能点三 安装与配置 JDK

1. JDK 的下载与安装

Java Development Kit 是 Sun 公司针对 Java 开发的软件开发工具包，简称为 JDK。JDK 作为 Java 的开发核心，包含了 Java 运行所需要的各种资源和工具，包括 Java 的运行环境、Java 编译器、常用的 Java 类等。

JDK 是 Java Web 开发的关键。安装 JDK 的步骤如下。

第一步：打开 JDK 官网（地址为 https://www.oracle.com/java/technologies/javase/javase-jdk8-downloads.html），下载 Windows 版本的 JDK，如图 1-8 所示。

Solaris SPARC 64-bit (SVR4 package)	125.19 MB	jdk-8u251-solaris-sparcv9.tar.Z
Solaris SPARC 64-bit	88.16 MB	jdk-8u251-solaris-sparcv9.tar.gz
Solaris x64 (SVR4 package)	133.64 MB	jdk-8u251-solaris-x64.tar.Z
Solaris x64	91.9 MB	jdk-8u251-solaris-x64.tar.gz
Windows x86	201.17 MB	jdk-8u251-windows-i586.exe
Windows x64	211.54 MB	jdk-8u251-windows-x64.exe

图 1-8 JDK 下载

第二步：双击下载的安装包（图 1-9），进入安装向导（图 1-10）。

图 1-9　JDK 安装包

第三步：单击图 1-10 中的“下一步”。

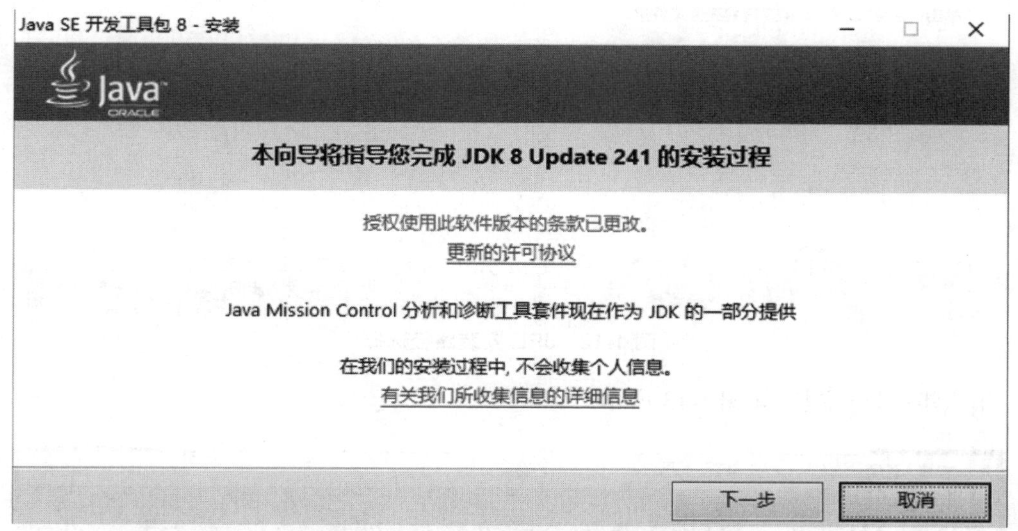

图 1-10　JDK 安装界面

第四步：默认会将程序安装在 C 盘，这里选择更改安装路径，单击“下一步”开始安装，如图 1-11 所示。

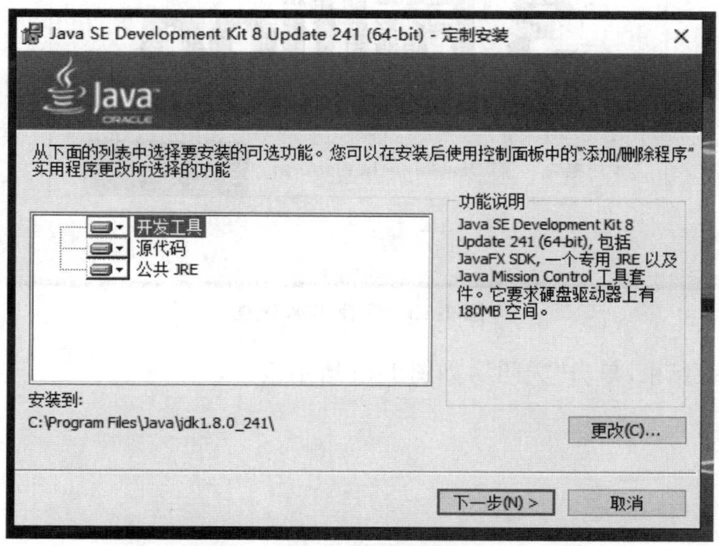

图 1-11　JDK 安装路径选择

第五步：安装 JRE，可以更改 JRE 的安装路径（与第四步选择相同的安装路径），如图 1-12 所示。

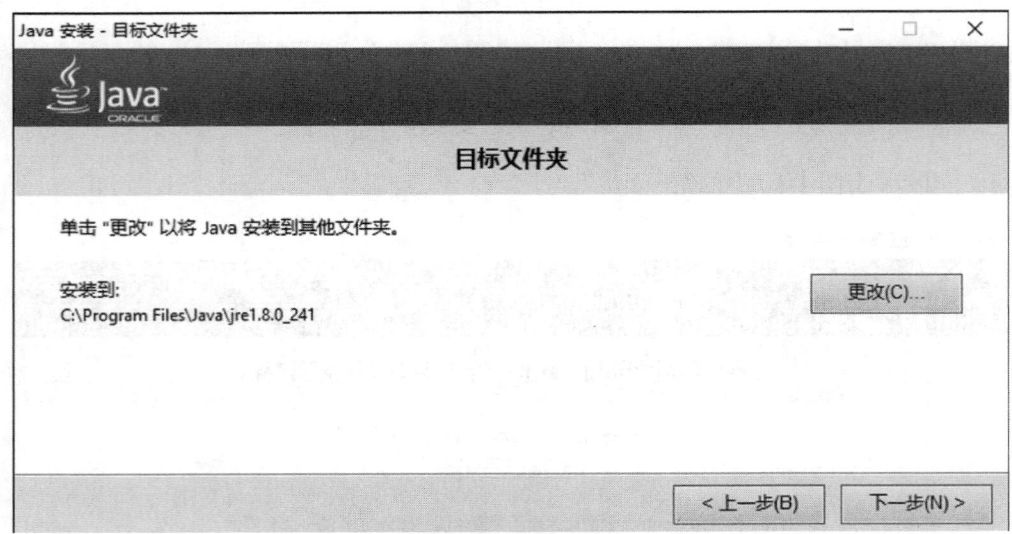

图 1-12 JRE 安装路径选择

第六步：进行安装，如图 1-13 所示。

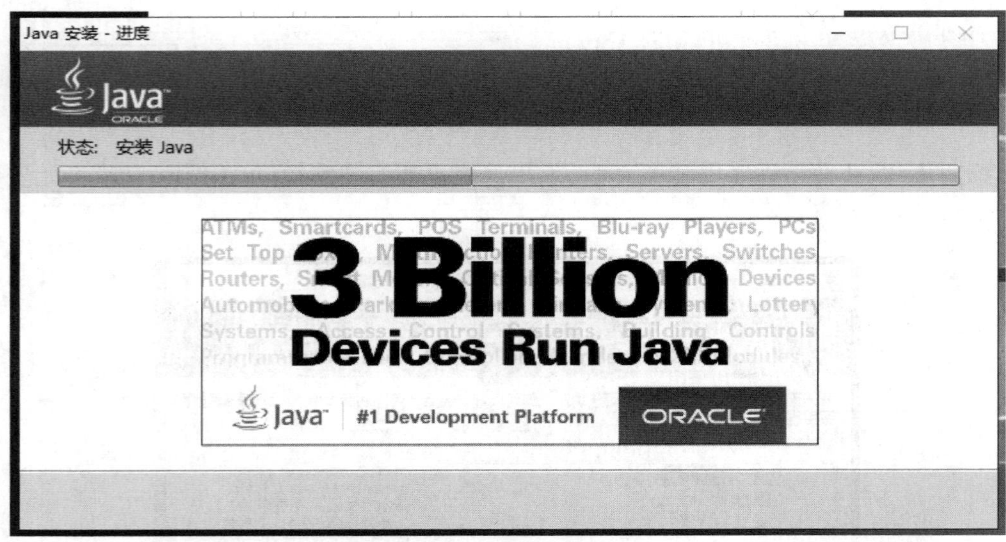

图 1-13 安装 JDK 过程

第七步：安装结束，单击“关闭”，如图 1-14 所示。

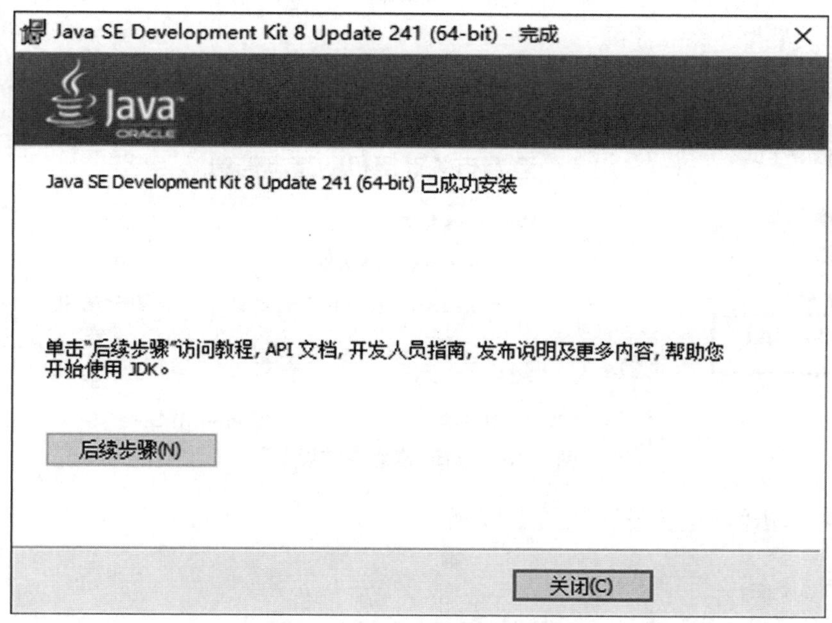

图 1-14　JDK 安装完成

2. 配置环境变量

安装 JDK 之后需要进行环境变量的配置，具体步骤如下。

第一步：打开“此电脑”，并在其上单击鼠标右键，选择“属性”，再选择“高级系统设置”，如图 1-15 和图 1-16 所示。

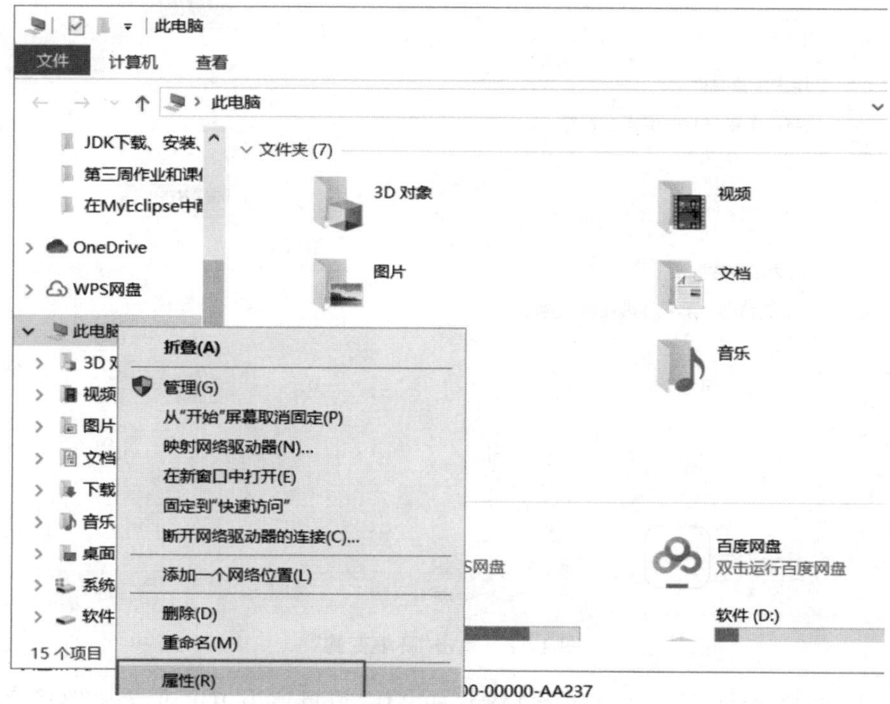

图 1-15　选择“属性”

系统

← → ⌄ ↑ > 控制面板 > 系统和安全 > 系统

控制面板主页

设备管理器

远程设置

系统保护

高级系统设置

查看有关计算机的基本信息

Windows 版本

Windows 10 专业版

© 2018 Microsoft Corporation。保留所有权利。

系统

处理器: Intel(R) Core(TM) i7-6500U CPU

图 1-16　选择“高级系统设置”

第二步：单击“环境变量”，如图 1-17 所示。

图 1-17　单击“环境变量”

第三步：新建系统变量，变量名为 JAVA_HOME，变量值为 JDK 的安装路径，如图 1-18

所示。

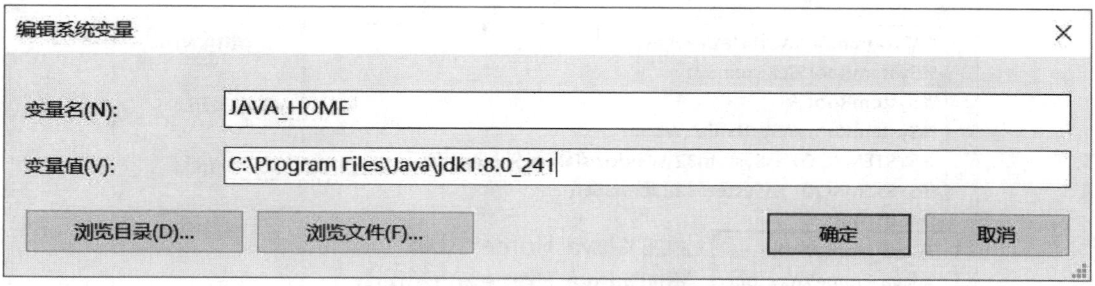

编辑系统变量
变量名(N): JAVA_HOME
变量值(V): C:\Program Files\Java\jdk1.8.0_241
浏览目录(D)...　浏览文件(F)...　确定　取消

图 1-18　配置 JAVA_HOME 系统变量

第四步：新建系统变量，变量名为 CLASSPATH，变量值为 .;%JAVA_HOME%\lib\dt.jar;%JAVA_HOME%\lib\tools.jar，如图 1-19 所示。

新建系统变量
变量名(N): CLASSPATH
变量值(V): .;%JAVA_HOME%\lib\dt.jar;%JAVA_HOME%\lib\tools.jar
浏览目录(D)...　浏览文件(F)...　确定　取消

图 1-19　配置 CLASSPATH 系统变量

第五步：找到 Path 变量并双击选中，单击“新建”(图 1-20)，设置变量值为 %Java_Home%\bin 以及 %Java_Home%\jre\bin，如图 1-21 所示。

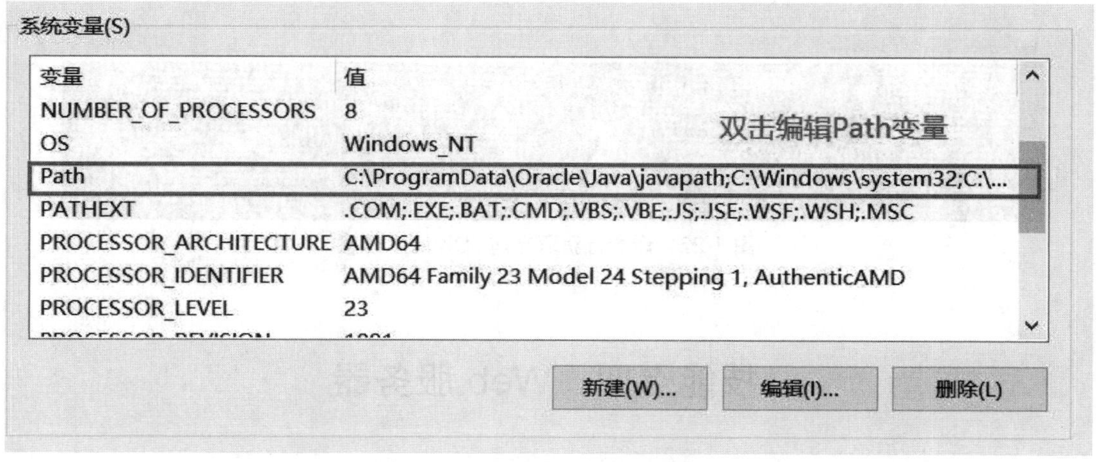

图 1-20　配置 Path 系统变量 1

图 1-21　配置 Path 系统变量 2

第六步：检验 JDK 是否配置成功。打开“运行”并输入 cmd 指令，用鼠标左键单击“确定”进入命令行窗口，输入 java -version 获取当前安装的 JDK 的版本信息，如图 1-22 所示。

```
C:\Windows\system32\cmd.exe
Microsoft Windows [版本 10.0.18362.836]
(c) 2019 Microsoft Corporation。保留所有权利。

C:\Users\11727>java -version
java version "1.8.0_25"
Java(TM) SE Runtime Environment (build 1.8.0_25-b18)
Java HotSpot(TM) 64-Bit Server VM (build 25.25-b02, mixed mode)

C:\Users\11727>
```

图 1-22　命令行获取当前 JDK 版本信息

技能点四　Web 服务器

1.Web 服务器概述

Web 服务器一般指网站服务器，是指存在于网络上某种类型计算机的程序，可以处理浏览器 Web 客户端的请求，并根据请求作出相应的响应；将资料资源放置在服务器上，以供所有访问服务器的用户下载使用。目前主流的三个 Web 服务器是 Nginx、IIS、Tomcat。

（1）Nginx

Nginx 是一款轻量级的 Web 服务器，其特点是并发能力强，占用内存少，稳定性强，具有丰富的功能集，系统资源消耗少。中国运用 Nginx 服务器的公司有百度、京东、新浪、网易等。Nginx 的图示如图 1-23 所示。

图 1-23　Nginx 图示

（2）IIS

IIS 是由微软公司开发的，基于运行 Windows 的 Web 服务器，其中包括 FTP 服务器、SMTP 服务器、NNTP 服务器和 Web 服务器，分别用于文件传输、邮件发送、新闻服务和页面浏览等方面。IIS 每次的更新迭代都会添加更多的功能，使安全性得到提升，并可实现多核扩展。IIS 的图示如图 1-24 所示。

图 1-24　IIS 图示

（3）Tomcat

Tomcat 属于轻量级应用服务器，是 Apache 软件基金会中的一个核心项目，因其免费开源、性能稳定且部署过程简单，被广大 Java 爱好者和开发商所认可，成为目前比较流行的 Web 应用服务器。本书案例使用 Tomcat 作为项目服务器，故其是重点学习的内容。Tomcat 的图示如图 1-25 所示。

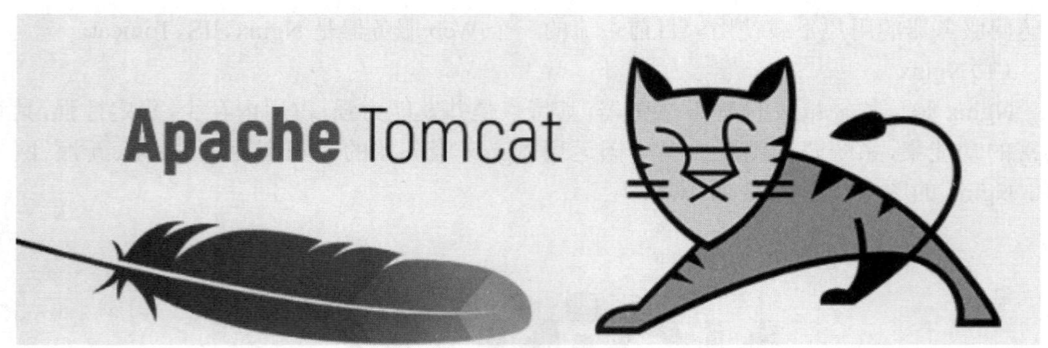

图 1-25 Tomcat 图示

2. Tomcat 的发展历史

Tomcat 是基于 Java 的 Web 应用服务器，实现了对 Servlet 和 JSP 的支持。由于 Tomcat 本身也存在一个 HTTP 服务器，因此可以被视作一个单独的 Web 服务器使用。Tomcat 运行时占用的系统资源少，支持负载平衡和邮件服务，扩展性好，并在不断改进和完善，故是开发和调试 JSP 程序的首选。经过多年的发展， Tomcat 每次更新都会提升自身的性能，这里主要介绍 Tomcat 5.x 至 Tomcat 9.x 版本所更新的重要内容 , 如表 1-2 所示。

表 1-2 Tomcat 版本内容

Tomcat 版本	说明
Tomcat 5.x	相比 Tomcat 之前的版本，新版本对底层代码进行了大量修改，提升了整体稳定性，优化了性能，减少了垃圾回收的动作，重构了程序部署，使用 Web 程序管理，提高了 Taglibs 的支撑能力，并集成了 Session 集群
Tomcat 6.x	吸取了 Tomcat 5.x 版本的优点，实现了 Servlet 2.5 和 JSP 2.1 的新特性，优化了内存的使用，拥有了更大的 I/O 容量
Tomcat 7.x	吸取了 Tomcat 6.x 的优点，实现了 Servlet 3.0 、JSP 2.0 和 EL 2.2 的新特性，优化了 Web 应用内存溢出的预防，增强了程序管理和服务器程序的安全性，可直接引用外部 Web 应用内容
Tomcat 8.x	支持 Java EE 7 规范，实现了 Servlet 3.1、JSP 2.3 和 EL 3.0 的新特性，对支持的资源进行了重构，可以更好地支持外部资源
Tomcat 9.x	新增了对于 HTTP/2.0 协议的支持以及 TLS 虚拟主机的支持，实现了 Servlet 4.0 规范草案

3.Tomcat 安装

网上书城项目选择 Tomcat 作为 Web 服务器。安装 Tomcat 的步骤如下。

第一步：用户可以通过 Tomcat 的主页（地址为 https://tomcat.apache.org/）中的下载链接进入 Tomcat 的下载页面，如图 1-26 所示。

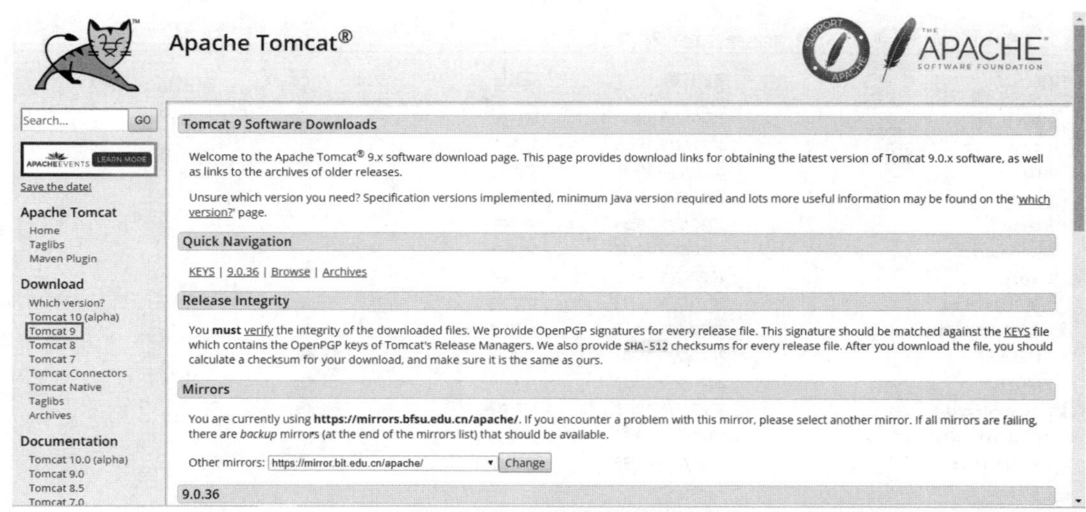

图 1-26　Tomcat 官方页面

第二步：选择下载版本 Tomcat 9.x，进入如图 1-27 所示的下载页面，zip 为压缩包的免安装版，installer 为安装版，根据 Windows 环境下载对应的程序包，这里选择 64 位压缩包免安装版进行下载。

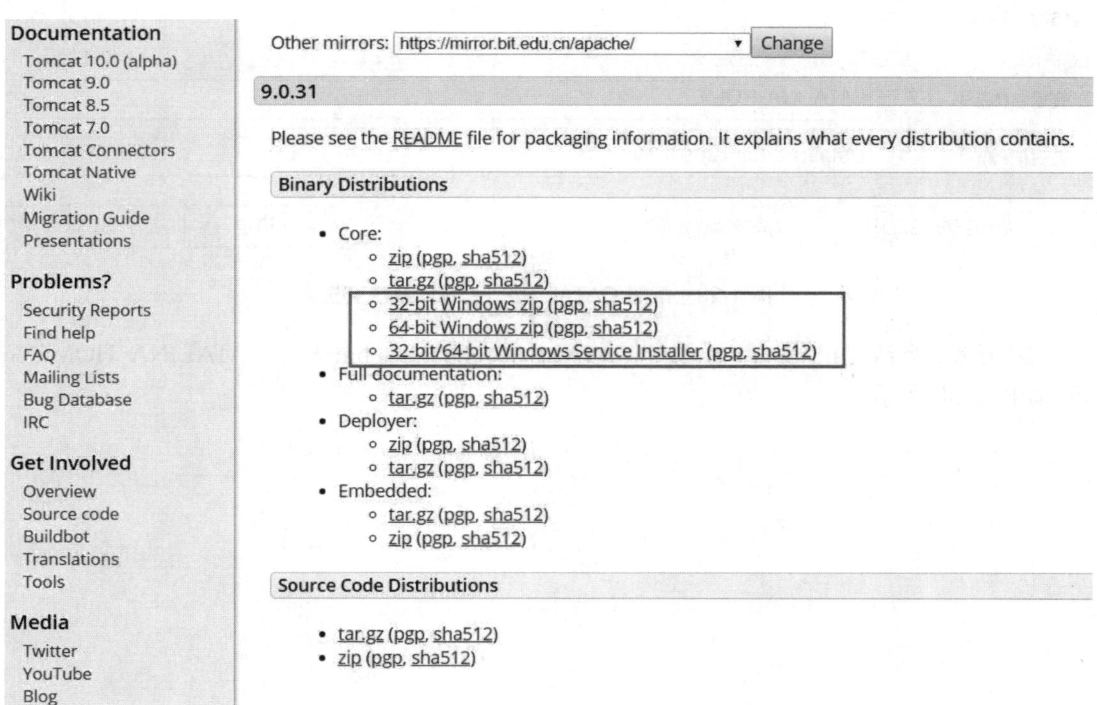

图 1-27　Tomcat 下载页面

第三步：下载压缩版 apache-tomcat-9.0.31-windows-x64.zip 并进行解压，解压后的文件内容如图 1-28 所示。

› 此电脑 › Windows8_OS (C:) › apache-tomcat-9.0.31

名称	修改日期	类型	大小
bin	2020/2/5 19:34	文件夹	
conf	2020/2/5 19:33	文件夹	
lib	2020/2/5 19:33	文件夹	
logs	2020/2/5 19:32	文件夹	
temp	2020/2/5 19:34	文件夹	
webapps	2020/2/5 19:34	文件夹	
work	2020/2/5 19:32	文件夹	
BUILDING.txt	2020/2/5 19:33	文本文档	20 KB
CONTRIBUTING.md	2020/2/5 19:33	MD文档	6 KB
LICENSE	2020/2/5 19:33	文件	57 KB
NOTICE	2020/2/5 19:33	文件	3 KB
README.md	2020/2/5 19:33	MD文档	4 KB
RELEASE-NOTES	2020/2/5 19:33	文件	7 KB
RUNNING.txt	2020/2/5 19:33	文本文档	17 KB

图 1-28 Tomcat 文件内容

第四步：打开电脑中的系统变量进行修改，新增变量 CATALINA_HOME，变量值为 Tomcat 解压文件夹，例如 c:\apache-tomcat-9.0.31，如图 1-29 所示。

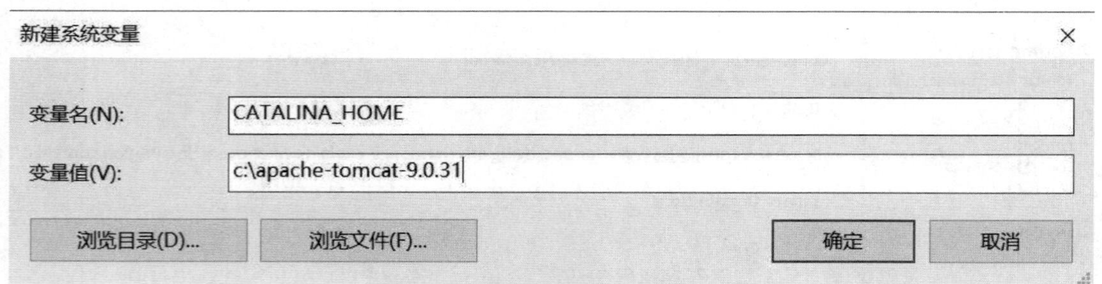

图 1-29 配置 CATALINA_HOME 系统变量

第五步：修改 Path 变量，添加 %CATALINA_HOME%\bin 和 %CATALINA_HOME%\lib，如图 1-30 所示。

图 1-30　配置 Path 环境变量

第六步：Tomcat 配置完成，现在要验证是否配置成功，在 Tomcat 目录中的 bin 目录中，运行命令行，输入 service.bat install，如图 1-31 所示。

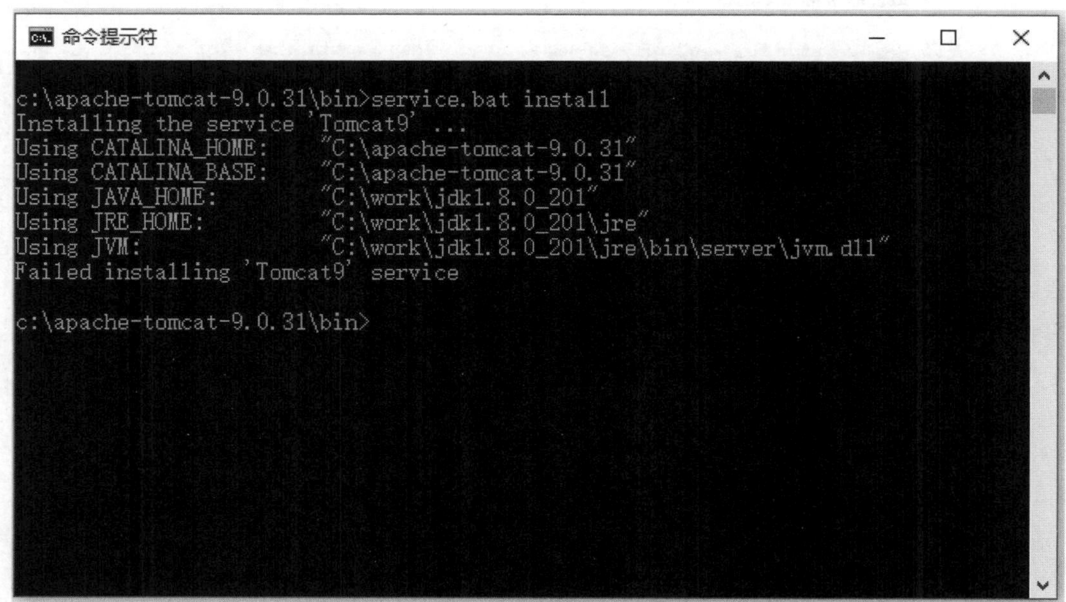

图 1-31　验证 Tomcat 是否安装成功

第七步：在命令行中输入 catalina start，浏览器输入 localhost:8080 并访问，如图 1-32 所示。

```
命令提示符

c:\apache-tomcat-9.0.31\bin>service.bat install
Installing the service 'Tomcat9' ...
Using CATALINA_HOME:    "C:\apache-tomcat-9.0.31"
Using CATALINA_BASE:    "C:\apache-tomcat-9.0.31"
Using JAVA_HOME:        "C:\work\jdk1.8.0_201"
Using JRE_HOME:         "C:\work\jdk1.8.0_201\jre"
Using JVM:              "C:\work\jdk1.8.0_201\jre\bin\server\jvm.dll"
Failed installing 'Tomcat9' service

c:\apache-tomcat-9.0.31\bin>catalina start
Using CATALINA_BASE:   "C:\apache-tomcat-9.0.31"
Using CATALINA_HOME:   "C:\apache-tomcat-9.0.31"
Using CATALINA_TMPDIR: "C:\apache-tomcat-9.0.31\temp"
Using JRE_HOME:        "C:\work\jdk1.8.0_201"
Using CLASSPATH:       "C:\apache-tomcat-9.0.31\bin\bootstrap.jar;C:\apache-tomc
at-9.0.31\bin\tomcat-juli.jar"

c:\apache-tomcat-9.0.31\bin>
```

图 1-32　Tomcat catalina 启动

第八步：配置成功后，访问该地址会显示 Tomcat 页面，如图 1-33 所示。

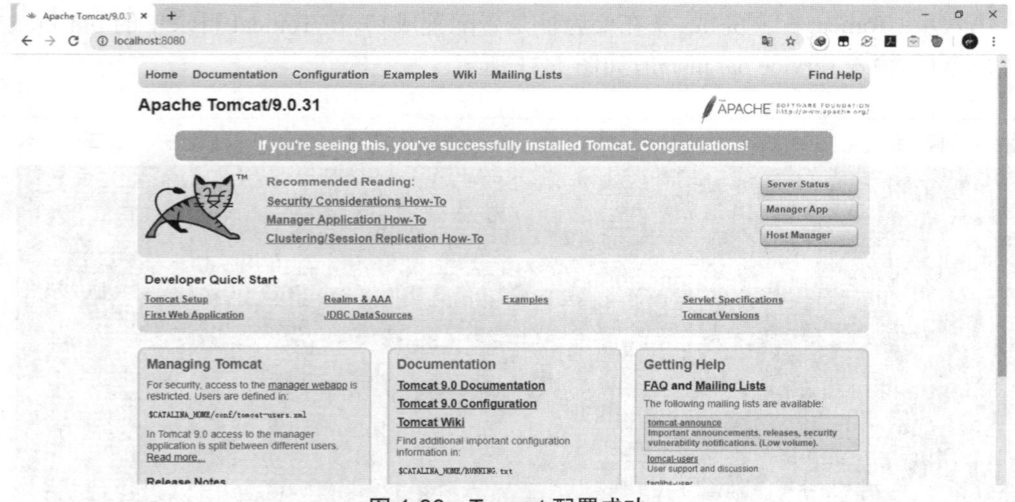

图 1-33　Tomcat 配置成功

技能点五　开发环境搭建

1. Eclipse

Eclipse 是开源代码，拥有可扩展的集成开发环境，它不仅可以用于 Java 桌面程序的开发，通过安装开发插件，还可以构建 Web 项目以及其他项目的开发环境。Eclipse 的图示如图 1-34 所示。

图 1-34　Eclipse 图示

Eclipse 作为一款开源开发工具，有如下特点：

- 开源免费使用；
- 非常适合 Java 语言的开发、编写、编译；
- 更新速度快；
- 插件开源免费使用，功能强大；
- 对于大型项目，在编译运行时会消耗大量系统资源；
- 插件的使用对于 Eclipse 版本要求严格，会出现插件更新的速度慢于软件更新的速度的情况；
- 对于“所见即所得”的 GUI 和 Web 界面设计，没有很好的支持。

2.IDEA

IntelliJ IDEA 简称 IDEA，是 Java 编程语言开发的集成环境。IDEA 对于代码自动补全、插件的应用以及重构，各种版本工具的支持都是非常优秀的。IDEA 可供选择的下载版本有两种：社区版（Community）和旗舰版（Ultimate）。社区版是免费的、开源的，但是功能较少；旗舰版提供了更多的功能，可免费试用 30 天，之后可使用激活码激活使用权限或是通过购买获得使用权限。IDEA 的图示如图 1-35 所示。

图 1-35　IDEA 图示

IDEA 作为一款优秀的开发软件，有很多特色功能，具体包括：

- 拥有丰富的导航模式；
- 在选取某个方法时，可从一个变量慢慢扩展到整个类，方便快捷；
- 可查看任何工程中的文件历史记录，方便恢复版本；
- 对于 Java 规范中提倡的方法可自动生成；
- 在调试模式下，可以对 Java 代码、JQuery、JavaScript 等技术进行调试。

安装 IDEA 的步骤如下。

第一步：登录 IDEA 官方网站（地址为 https://www.jetbrains.com/idea），如图 1-36 所示。

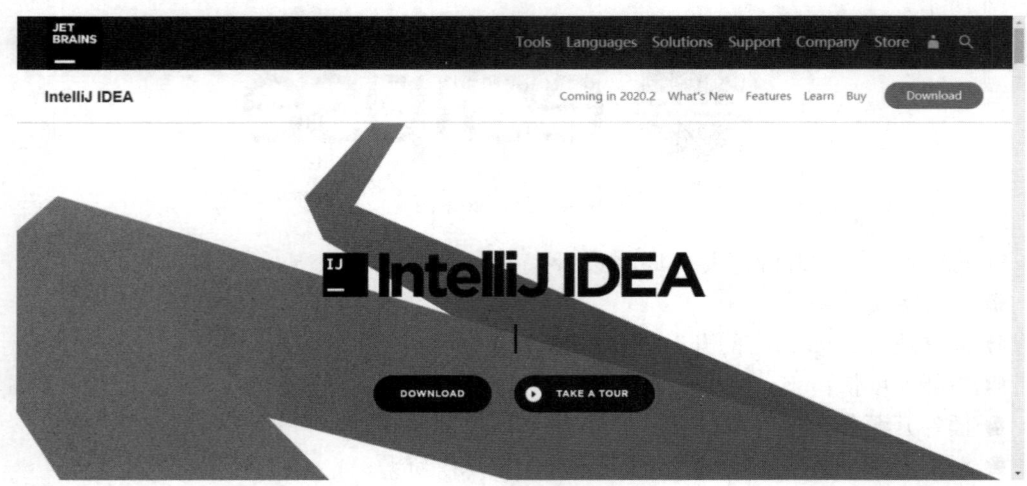

图 1-36 IDEA 官方网站

第二步：单击“DOWNLOAD”，进入如图 1-37 所示的界面。

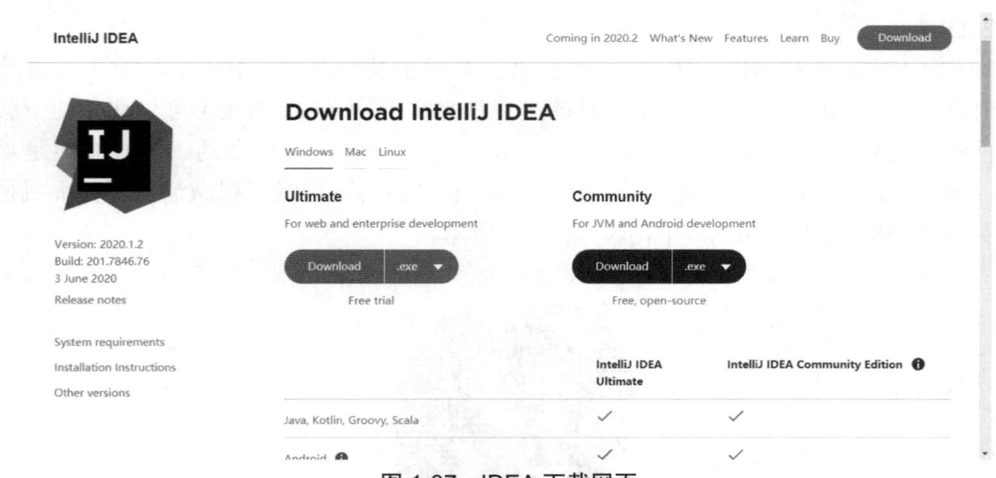

图 1-37 IDEA 下载网页

第三步：选择旗舰版“Ultimate”并下载，下载完成之后双击可执行文件开始安装，在如图 1-38 所示的对话框中单击“Next”。

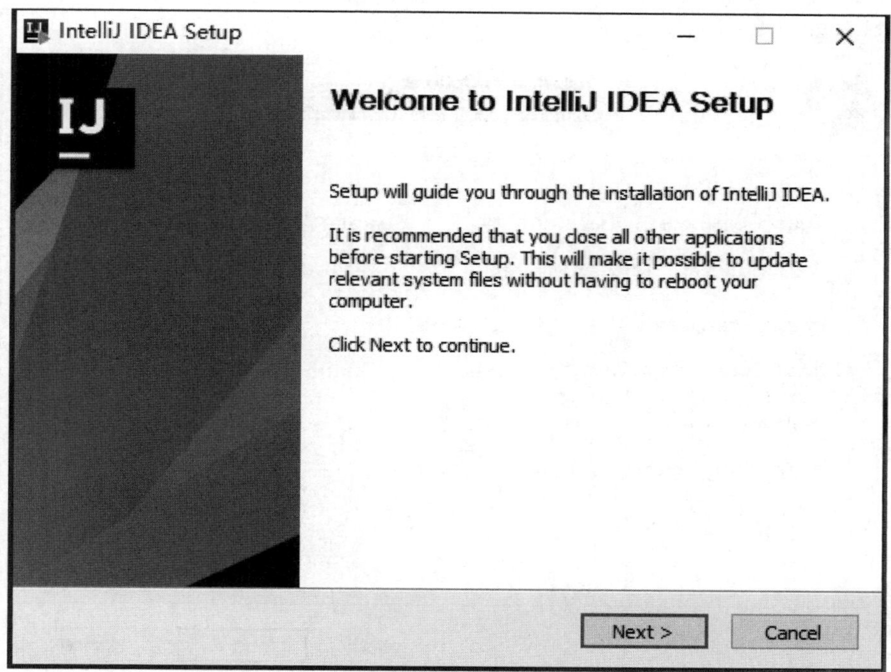

图 1-38　IDEA 安装

第四步：选择安装的路径，单击“Next”，如图 1-39 所示。

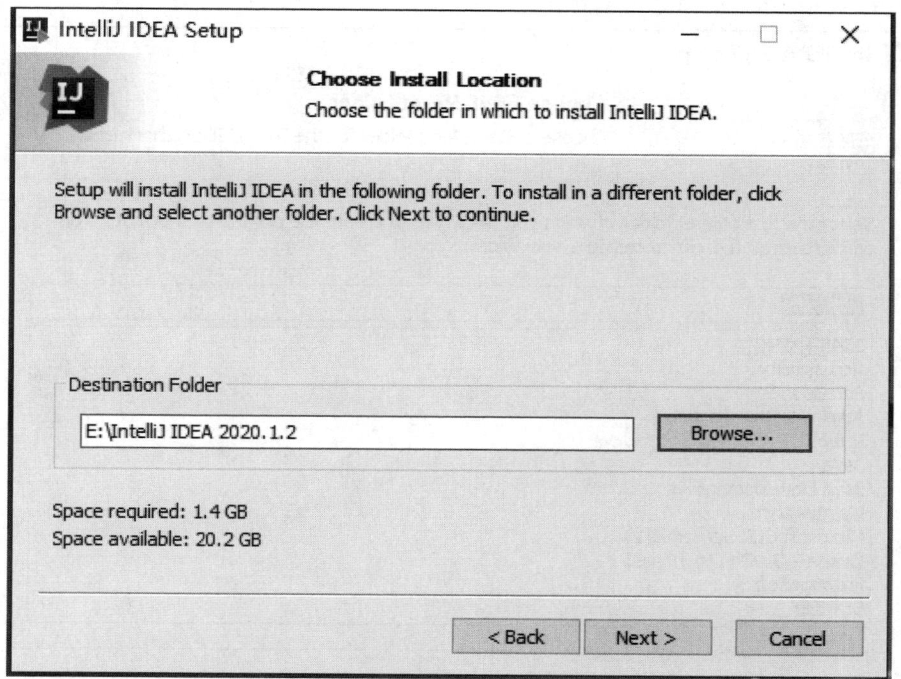

图 1-39　选择安装路径界面

第五步：选择与设备对应的参数，并选择语言为“.java”，如图 1-40 所示。

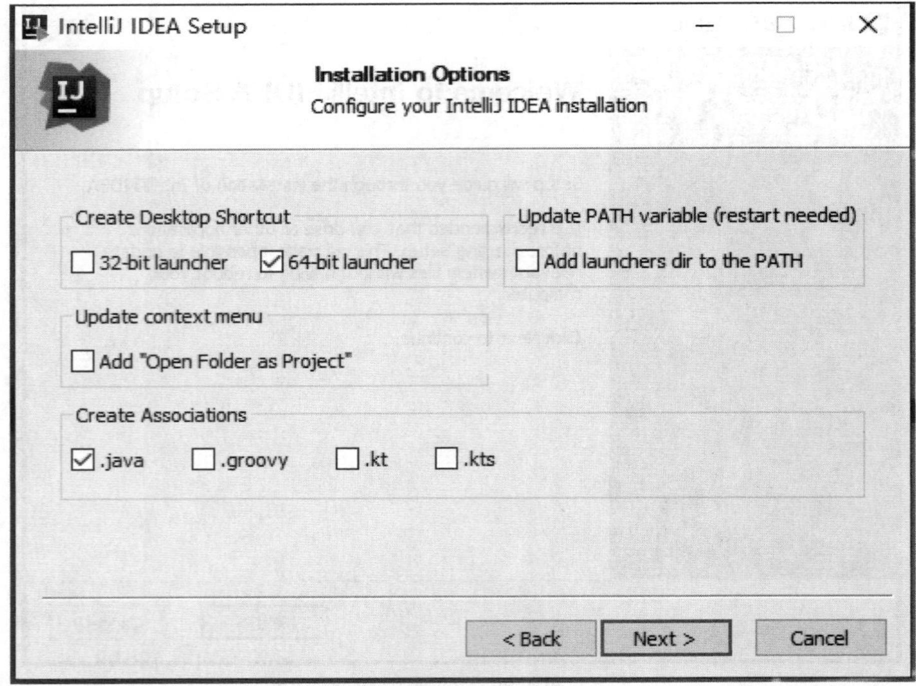

图 1-40 选择安装设备参数

第六步：选择开始菜单文件夹并单击“Install”，开始安装，如图 1-41 所示。

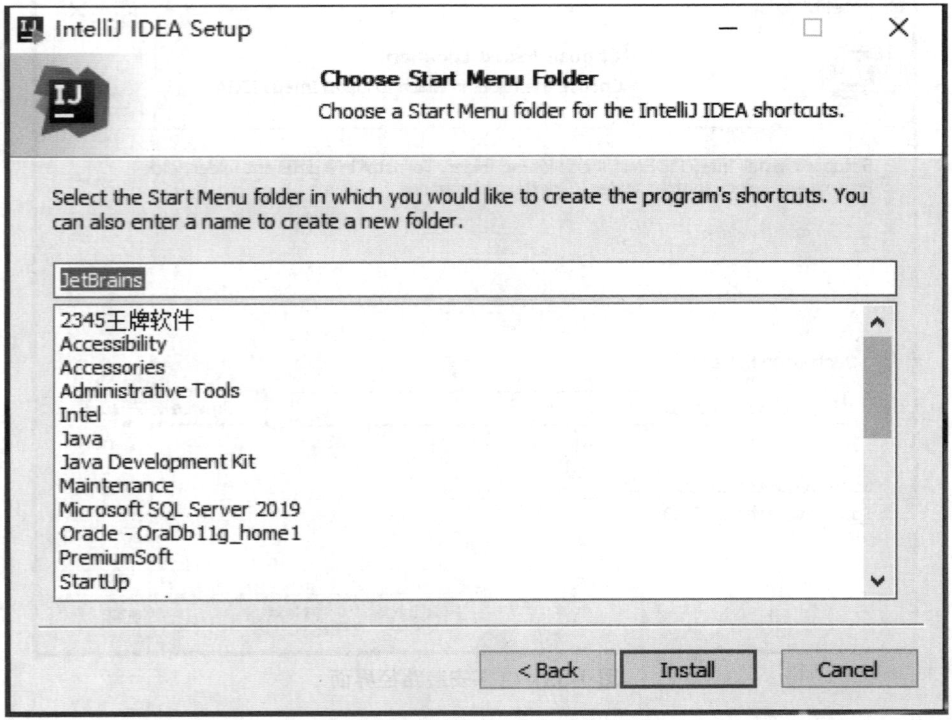

图 1-41 选择开始菜单文件夹

第七步：安装完成，如图 1-42 所示。

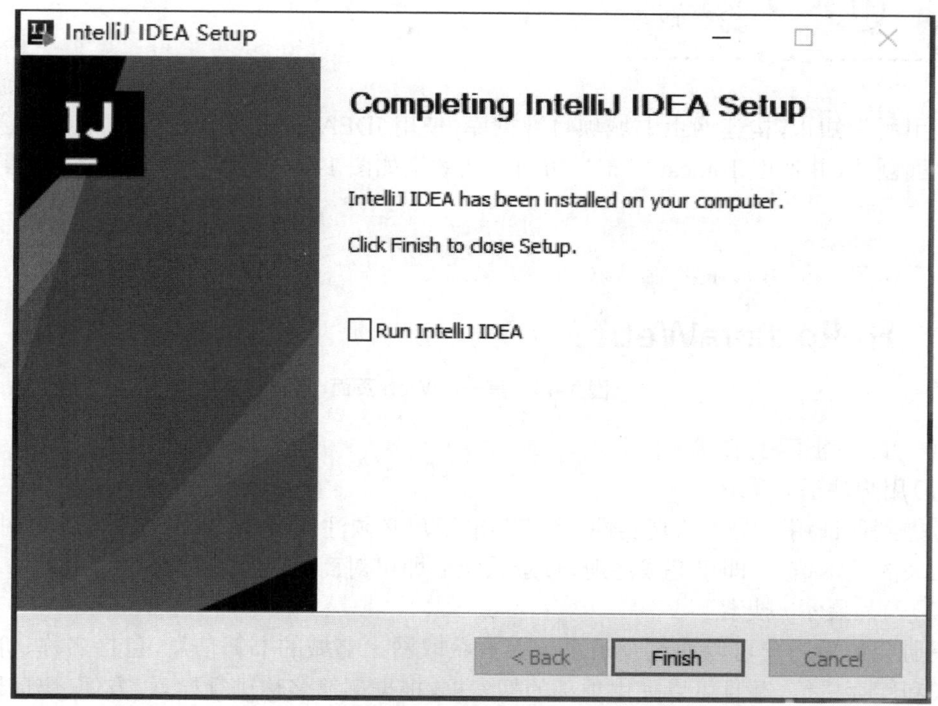

图 1-42　安装完成

第八步：打开 IDEA 软件，如图 1-43 所示。

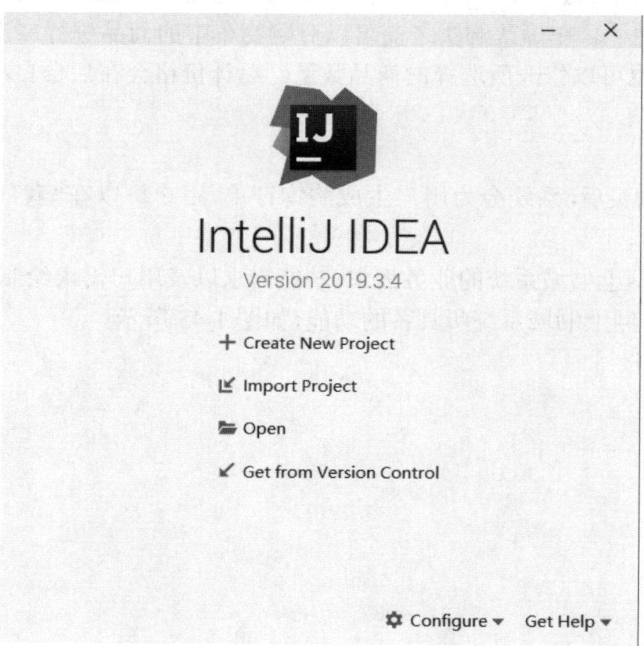

图 1-43　IDEA 界面

运用所学知识，构建网上书城项目数据库，使用 IDEA 创建网上书城项目，完成第一个 Web 页面创建，并使用 Tomcat 发布且访问。其效果如图 1-44 所示。具体的操作步骤如下。

图 1-44 第一个 Web 页面

第一步：分析网上书城项目需求。

（1）用户注册 / 登录

考虑到保证用户购买的真实性，系统要求用户必须注册会员，才能登录。注册时，用户填写相关的基本信息，即可完成注册，登录系统后即可对商品进行选购。

（2）商品展示 / 搜索

注册会员后用户可以通过登录系统首页获取网上书城的书籍信息，包括名称、价格、书籍样式和库存量等，并且可以使用顶部的搜索框，根据书籍名称进行检索，方便、快捷地查询自己感兴趣的书籍。

（3）购物车

用户在浏览商品的过程中，对于想要购买的书籍可进行勾选，在选择完成后需单击右下方的“添加到购物车”。在确认购买之前可以对购物车中的商品进行二次选择，既可以从购物车中删除商品，也可以修改所选择的商品数量。总计价格会在后台自动生成，并在前端实时显示。

（4）订单

用户在确认购买后，系统会为用户生成购物订单，用户可以在“我的订单”中查看自己的订单信息。

第二步：根据网上书城系统的业务需求、功能需求以及用户需求绘制网上书城系统的用例图。该图描述了网上书城系统所具备的功能，如图 1-45 所示。

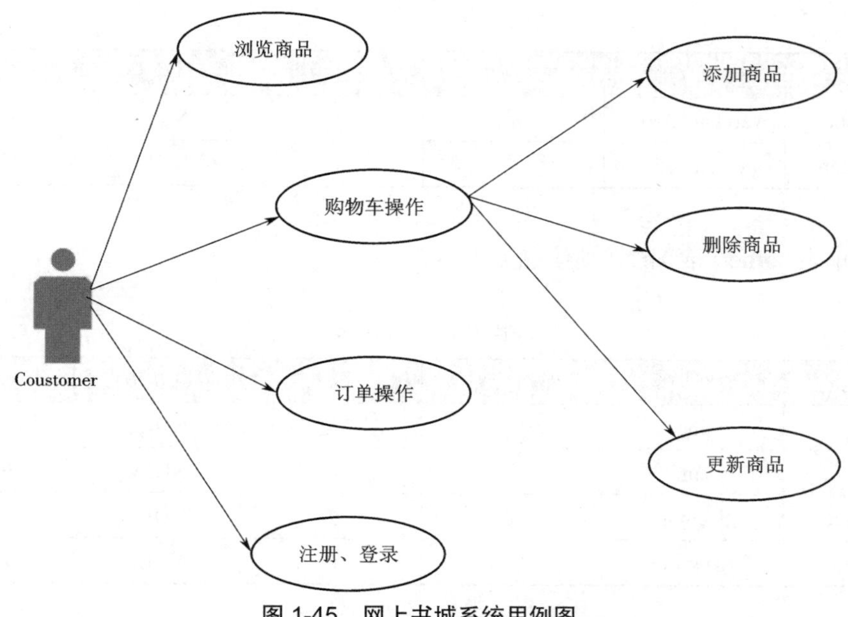

图 1-45　网上书城系统用例图

第三步：根据系统功能描述和实际业务分析，对网上书城系统进行数据库设计，其主要的数据表如下。

1）用户信息表（userinfo 表）的详细信息见表 1-3。

表 1-3　userinfo 表

字段名	数据类型	是否为空	是否主键	默认值	描述
uid	int	否	是	NULL	系统自增编号
username	varchar（50）	否	否	NULL	用户名
password	varchar（50）	否	否	NULL	用户密码
email	varchar（255）	是	否	NULL	电子邮箱

2）书籍表（books 表）的详细信息见表 1-4。

表 1-4　books 表

字段名	数据类型	是否为空	是否主键	默认值	描述
bid	int	否	是	NULL	系统自增编号
bookname	varchar（100）	是	否	NULL	书籍名称
price	decimal	是	否	NULL	价格
image	varchar（200）	是	否	NULL	书籍图片
stock	int	是	否	NULL	库存

续表

字段名	数据类型	是否为空	是否主键	默认值	描述
booknumber	varchar(200)	是	否	NULL	书籍编号
introduction	varchar(255)	是	否	NULL	书籍简介

3)订单表(orders 表)的详细信息见表 1-5。

表 1-5 orders 表

字段名	数据类型	是否为空	是否主键	默认值	描述
oid	int	否	是	NULL	系统自增编号
uid	int	否	否	NULL	用户 ID
total_price	decimal	是	否	NULL	总价
createdate	datetime	是	否	NULL	生成订单日期

4)订单商品信息表(items 表)的详细信息见表 1-6。

表 1-6 items 表

字段名	数据类型	是否为空	是否主键	默认值	描述
iid	int	否	是	NULL	系统自增编号
oid	int	否	否	NULL	订单 ID
bid	int	否	否	NULL	书籍 ID
createdate	datetime	是	否	NULL	生成时间
count	int	是	否	NULL	订购书籍数量
deal_price	decimal	是	否	NULL	总价
state	int	是	否	NULL	支付状态

第四步:打开 IDEA,单击“Create New Project”创建一个新项目,如图 1-46 所示。

图 1-46　启动 IDEA

第五步：在界面中选择“Java Enterprise”→选择 SDK、Java EE 版本以及 Tomcat 版本→选择“Web Application（4.0）”→单击“Next”，即可创建 Java Web 项目，如图 1-47 所示。

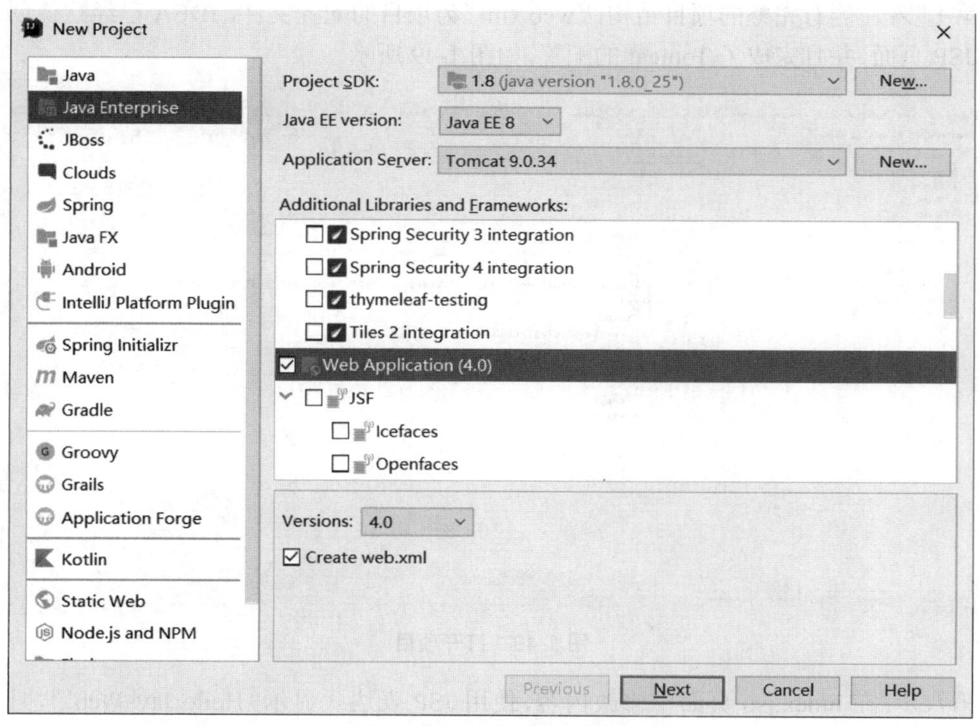

图 1-47　创建 Web 项目

第六步：修改新创建的项目名称和路径，单击“Finish”完成项目的创建，如图 1-48 所示。

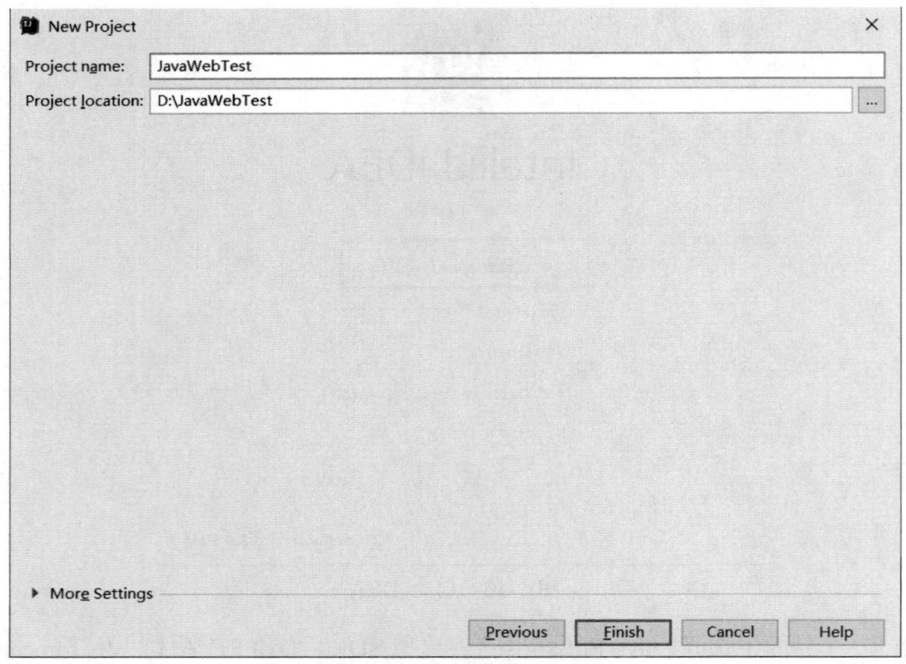

图 1-48 设置项目名称和路径

第七步：左侧有完整的项目结构，“web.xml”为项目的配置文件，IDEA 已经默认创建了一个 JSP 页面，并且完成了 Tomcat 的配置，如图 1-49 所示。

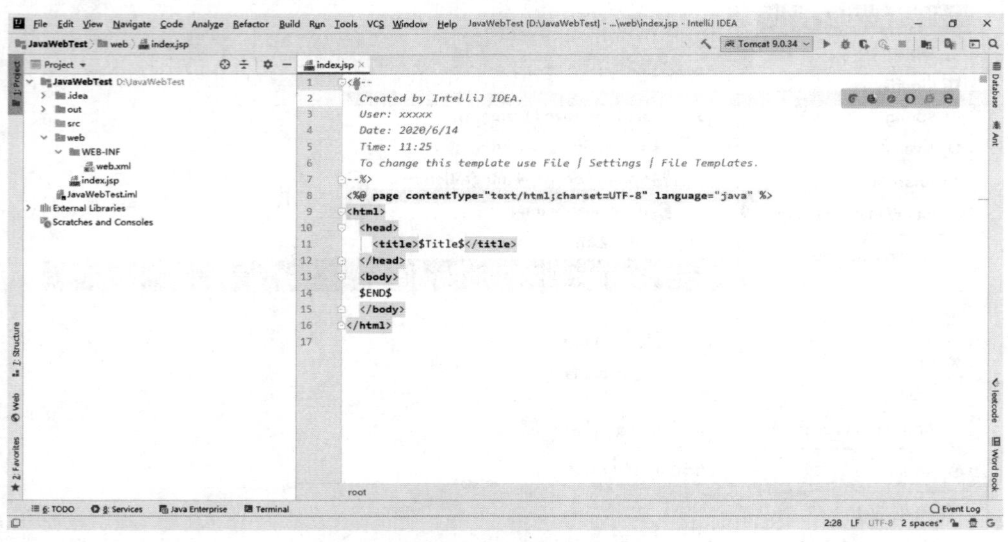

图 1-49 打开项目

第八步：在 index.jsp 页面中修改内容，使用 JSP 表达式显示“Hello JavaWeb!”，具体如示例代码 1-1 所示。

示例代码 1-1：index.jsp

```
<%@ page contentType="text/html;charset=UTF-8" language="java" %>
<html>
<head>
    <title>JavaWeb</title>
</head>
<body>
<h1><%="Hello JavaWeb!"%></h1>// 使用 JSP 表达式将“Hello JavaWeb!”字符串输出到页面上
</body>
</html>
```

若要修改 Tomcat 配置可以单击右上角对应的“Edit Configurations”，修改 Tomcat 配置信息，如图 1-50 所示。

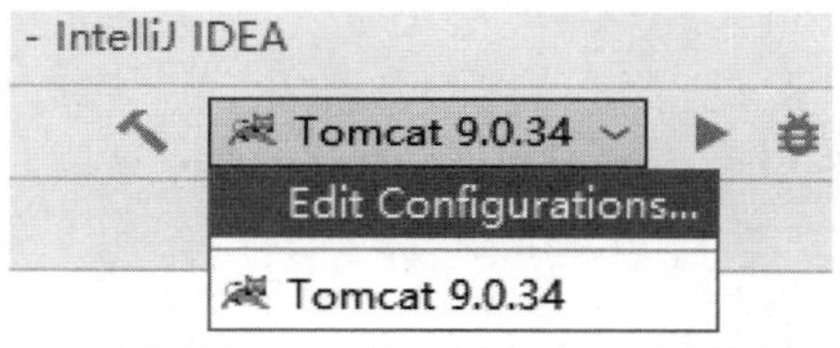

图 1-50　“Edit Configurations”位置

在“Edit Configurations”配置页面，可以修改地址、Tomcat 版本、端口号等配置，如图 1-51 所示。

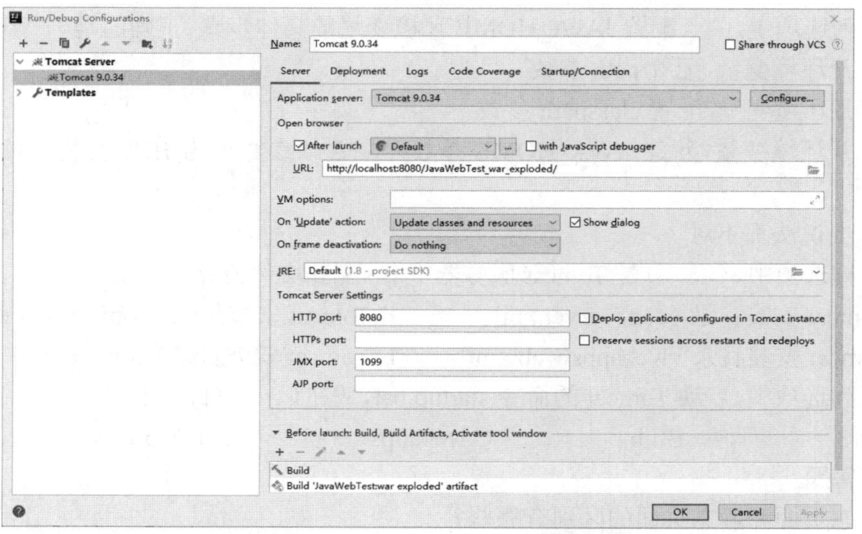

图 1-51　修改 IDEA 中 Tomcat 配置

本次任务通过对 Java Web 基本知识的介绍和网上书城项目环境的搭建，使读者了解 C/S 模式和 B/S 模式的不同，掌握静态页面和动态页面的构成，掌握基本 Web 项目的搭建方法，能够通过 IDEA 搭建 Web 项目，并运行第一个 Web 项目页面。

browser	浏览器	client	客户机
service	服务	development	发展
version	版本	ultimate	终极的

1. 选择题

1）如果做动态网站开发，以下（　　）可以作为服务器端脚本语言。

A.Java　　B.JSP　　C.JavaScript　　D.HTML

2）以下文件名后缀中，只有（　　）是静态网页的后缀。

A.jsp　　B.html　　C.aspx　　D. php

3）下列选项中，关于配置 JAVA_HOME 环境变量的具体步骤，正确的是（　　）。

A. 打开环境变量，配置 path 参数

B. 打开环境变量，配置 classpath 参数

C. 打开环境变量，新建 JAVA_HOME 参数并配置该参数值为 JDK 安装时对应的安装目录路径

D. 以上说法都不对

4）下列选项中，（　　）是 Tomcat 服务器中 jsp 配置文件的存放目录。

A.Tomcat 安装目录 \ conf\server.xml　　B.Tomcat 安装目录 \ conf\web.xml

C.Tomcat 安装目录 \ webapps\web.xml　　D.Tomcat 安装目录 \ conf\context.xml

5）下列选项中，启动 Tomcat 的命令 startup.bat，放在（　　）目录中。

A.bin　　B.lib　　C.webapps　　D.work

2. 简答题

1）动态页面和静态页面的区别有哪些？

2）B/S 模式的优点有哪些？

第二章　网上书城项目登录页面设计

通过学习 JSP 相关知识，结合 HTML 基础知识，掌握 JSP 的基本语法、指令标记和动作标记的有关知识，具有运用所学的相关知识编写网上书城项目登录页面的能力。在任务实现过程中：

- 了解 JSP 基本概念和页面构成；
- 掌握 JSP 注释、声明、脚本的使用；
- 掌握 JSP 指令标记的使用；
- 掌握 JSP 动作标记的使用。

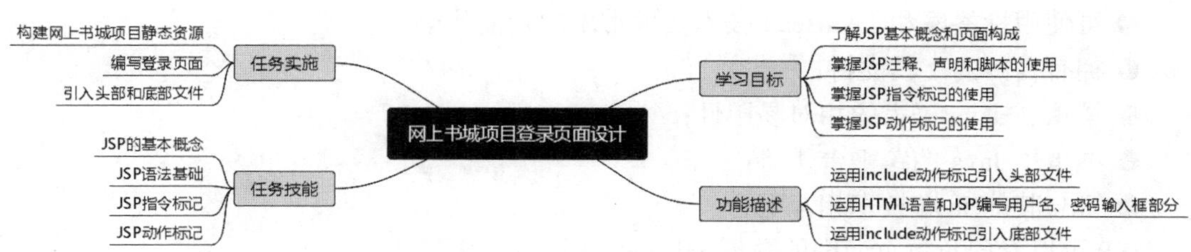

【情境导入】

在我们日常生活中，一些网站上会充斥着各种各样的网上资源，而对于网页内容数据的获取一般都是动态的，可是一般的 HTML 资源都是静态的，资源不会随着数据动态变化，例如当前的时间、登录的人员信息、历史记录等。本次任务就是使用 JSP 实现网上书城项目登录页面的设计。

【功能描述】

- 运用 include 动作标记引入头部文件。
- 运用 HTML 语言和 JSP 编写用户名、密码输入框部分。
- 运用 include 动作标记引入底部文件。

技能点一 JSP 的基本概念

1. JSP 基本概念

Java Server Pages 是由 Sun 公司创建的一种动态网页技术，简称 JSP。JSP 可部署在网络的服务器上，以此来响应用户发送的请求，并根据发送的请求配合 HTML 语言或其他格式的 Web 网页，动态响应用户需求，然后返回至浏览器，显示给用户。JSP 技术以 Java 作为主要语言，并能与服务器上的其他 Java 程序共同处理复杂的业务需求。

JSP 发展至今有以下诸多特点：

- 能通过模板化的方式简单、高效地添加动态网页内容；
- 可使用标签库和 JavaBean 技术实现常用的功能代码；
- 拥有良好的工具支持；
- 继承了 Java 语言的相对易用性；
- 继承了 Java 的跨平台优势；
- 可与其他 Java 技术相互配合。

JSP 可以专门负责页面中的数据呈现，实现分层开发。

2. JSP 基本原理

Web 项目运行时，用户通过单击 JSP 页面发送请求给服务器，此时 JSP 引擎会通过预处理把 JSP 文件中的静态数据（HTML 文本）和动态数据（Java 脚本）全部转换为 Java 代码。将生成的 .java 文件编译成 Servlet 类文件（.class），Tomcat 服务器生成的类文件默认存放在 \work 目录，最后将编译后的 class 对象加载到容器中，并根据用户的请求生成 HTML 格式的响应页面。

模拟用户首次访问 test.jsp 页面，JSP 编译的过程如图 2-1 所示。

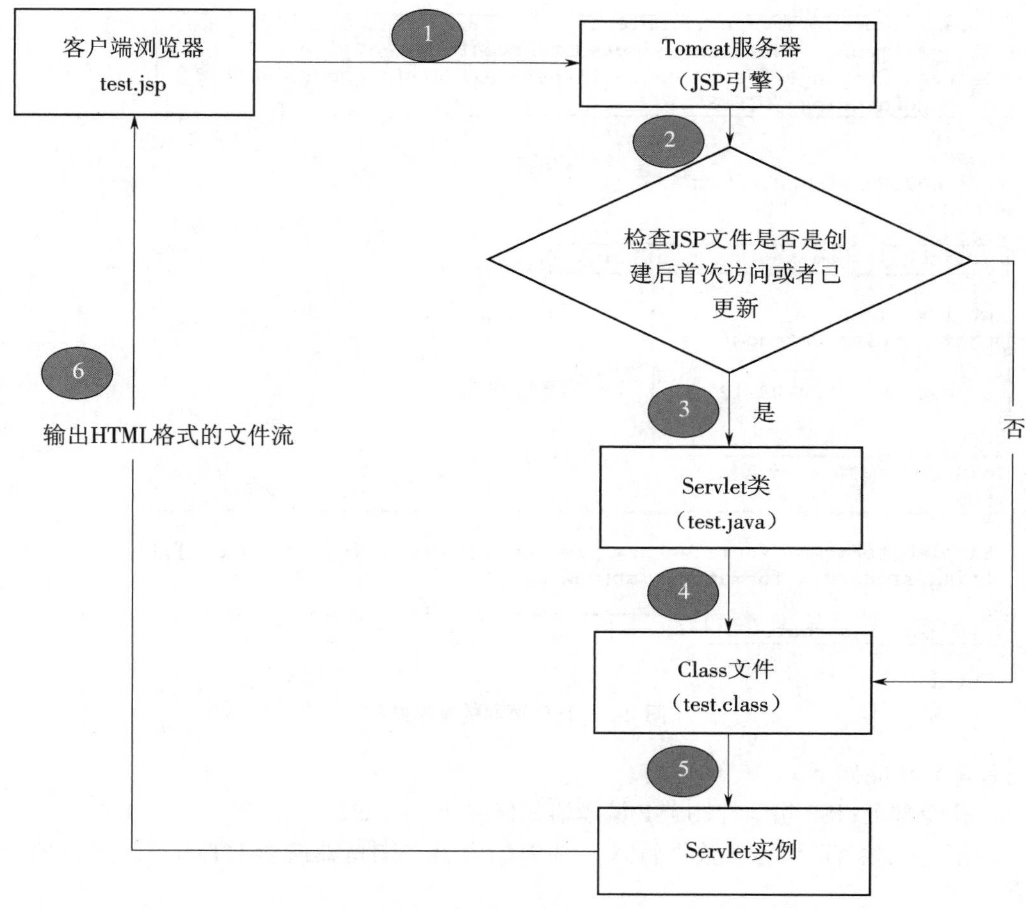

图 2-1　JSP 编译的过程

当客户端发出请求，且请求为 JSP 时，到相应的 Servlet 进行处理，test.jsp 转化为 test.java。再将 Servlet 转化为 test.class 文件，将 class 文件加载到容器中，并在容器中创建一个实例进行初始化。通过 Servlet 实例中的 JSP 中的 Service 方法，将 HTML 文件返回至客户端。

3. JSP 页面基本构成

JSP 页面是 Java 代码与传统 HTML 语言的相互结合，以达到动态显示数据和获取数据的目的。JSP 页面基本结构由指令部分、静态内容部分、注释部分、声明部分、脚本部分和表达式部分组成，如图 2-2 所示。

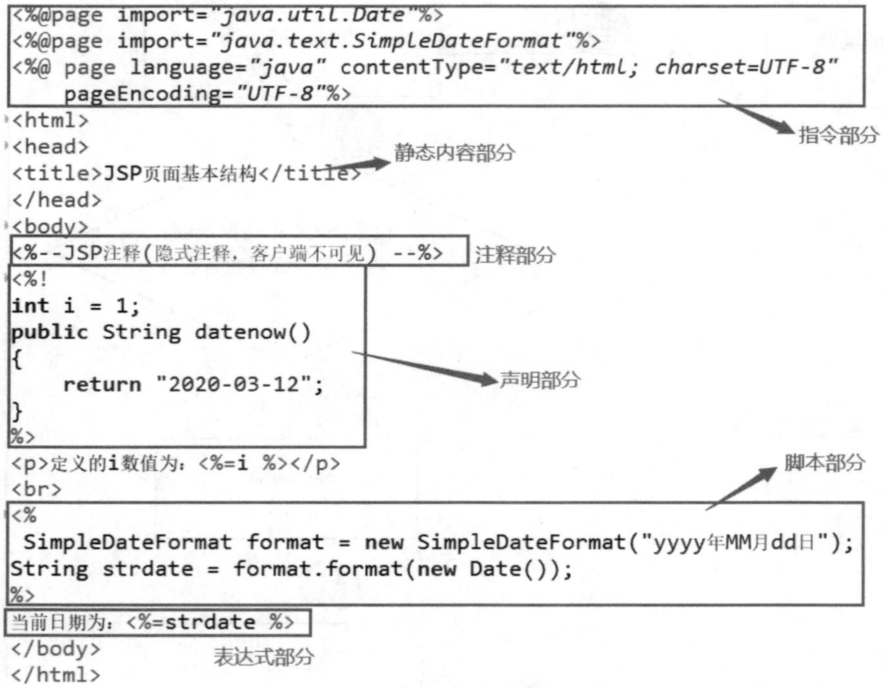

```
<%@page import="java.util.Date"%>
<%@page import="java.text.SimpleDateFormat"%>
<%@ page language="java" contentType="text/html; charset=UTF-8"
    pageEncoding="UTF-8"%>
<html>
<head>
<title>JSP页面基本结构</title>
</head>
<body>
<%--JSP注释(隐式注释, 客户端不可见) --%>
<%!
int i = 1;
public String datenow()
{
    return "2020-03-12";
}
%>
<p>定义的i数值为: <%=i %></p>
<br>
<%
 SimpleDateFormat format = new SimpleDateFormat("yyyy年MM月dd日");
String strdate = format.format(new Date());
%>
当前日期为: <%=strdate %>
</body>
</html>
```

指令部分
静态内容部分
注释部分
声明部分
脚本部分
表达式部分

图 2-2　JSP 页面的基本结构

各部分功能如下。

1）指令部分：JSP 指令控制 JSP 编译器如何生成 Servlet。

2）静态内容部分：静态数据输入文件中的内容在浏览器解析 HTML 标记后，展示其内容。

3）注释部分：JSP 专用的“<%-- 注释内容 --%>”注释方法为隐式注释，客户端不可见。

4）声明部分：使用“<%!...%>”将声明的变量、方法等编写在其中。

5）脚本部分：使用“<%...%>”将 Java 代码编写在其中，可以定义局部变量或者调用方法，但不能定义方法。

6）表达式部分：“<%=...%>”是 <% out.println（变量）%> 的简写方式，用于将已经声明的变量或者表达式输出到网页上。

与 JSP 指令元素不同的是，JSP 动作元素在请求处理阶段起作用。一个 JSP 页面中不一定包含所有元素，不过在复杂程度较高的项目中，会看到这些内容。

技能点二　JSP 的基本语法

在 JSP 页面的 HTML 中嵌入 Java 代码，这部分代码称为脚本（scripting）。脚本元素是 JSP 页面中的内嵌代码，通常用 Java 编程语言编写。

JSP 有几种不同类型的脚本：声明（declaration）、表达式（expression）、代码段

（scriptlet）、注释，具体内容如下。

1. 声明

声明元素用于在 JSP 页面中声明变量、常量和方法。声明以“<%!”标记开始，以“%>”标记结束，其中“%”与“!”必须紧挨，不能有空格。

1）声明变量的语法如下：

```
<%! 声明 1, 声明 2,…;%>
```

例如：在 JSP 中声明相关变量的代码如下。

```
<%! int i=0; %>// 声明单个变量
<%! int a,b,c;%>// 同时声明多个变量，变量用逗号分隔
<%! Integer obj=new Integer(10);%>// 声明对象
```

2）在声明的变量前加上关键词 final 即为声明常量。声明常量的语法如下：

```
<%! final 类型 = 值 ; %>
```

例如：声明一个常量字符串 name 值为 abc 的代码如下。

```
<%! final String name="abc";%>
```

3）声明方法的语法如下：

```
<%!
  public 返回类型 函数名（参数列表）{
  函数体 ;
}
%>
```

例如：声明一个方法，并返回一段字符串“Hello,javaweb”的代码如下。

```
<%!
  public String say(){// 定义公有方法 say，返回值为字符串类型
    return "Hello,javaweb";// 返回一段字符串
  }
%>
```

声明元素中声明的变量、常量和方法可以在 JSP 页面中由其他脚本元素、表达式或 JSP 动作使用。

例如：使用 JSP 定义变量 i、常量 name 和方法 say 并进行输出，效果如图 2-3 所示。

图 2-3　声明实例

一个 JSP 页面中可以有多个声明元素，JSP 容器在处理此页面时会把这些声明元素合并为一个，如示例代码 2-1 所示。

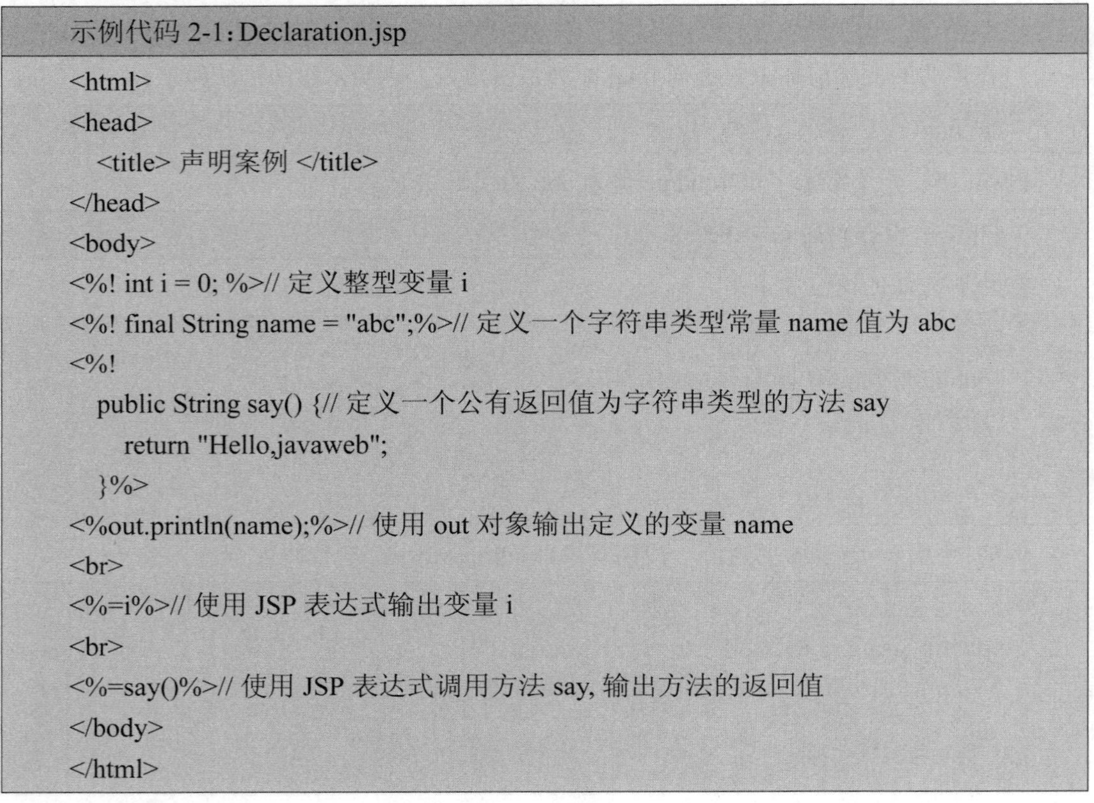

示例代码 2-1：Declaration.jsp

```
<html>
<head>
  <title> 声明案例 </title>
</head>
<body>
<%! int i = 0; %>// 定义整型变量 i
<%! final String name = "abc";%>// 定义一个字符串类型常量 name 值为 abc
<%!
  public String say() {// 定义一个公有返回值为字符串类型的方法 say
    return "Hello,javaweb";
  }%>
<%out.println(name);%>// 使用 out 对象输出定义的变量 name
<br>
<%=i%>// 使用 JSP 表达式输出变量 i
<br>
<%=say()%>// 使用 JSP 表达式调用方法 say, 输出方法的返回值
</body>
</html>
```

2. 表达式

JSP 的表达式元素是一个内嵌的 Java 表达式，这个表达式经过运算执行会得到一个结果字符串。此字符串可通过输出表达式将结果在客户端浏览器中进行显示。输出表达式的语法格式如下：

```
<%= 表达式 %>
```

表达式元素有以下特点：

1）表达式元素都用“<%=”和“%>”括起；

2）表达式元素中包含由 Java 编程语言编写的表达式，可以是一个变量、一个字段或者是一个方法调用的结果；

3）输出内容 / 输出结果会嵌入 HTML 中。

例如：使用表达式元素将字符串 sayHello 以及方法 say 返回值显示到页面，效果如图 2-4 所示。

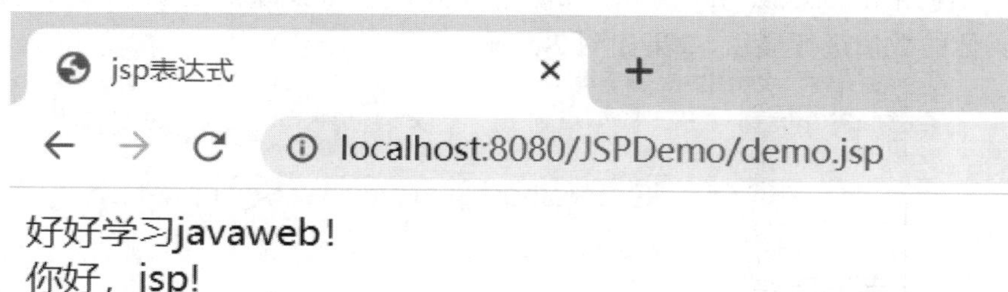

图 2-4　JSP 表达式实例

实现图 2-4 所示的效果，如示例代码 2-2 所示。

示例代码 2-2：demo.jsp

```
<%@page language="java" import="java.util.*" pageEncoding="GBK" %>
<%!
  String sayHello = " 好好学习 javaweb！ ";// 定义字符串类型变量 sayHello

  public String say() {// 定义公有类型名为 say 的方法，其返回值为字符串类型
    return " 你好，jsp!";// 返回一个字符串“你好，jsp!”
  }
%>
<html>
<head>
  <title>JSP 表达式 </title>
</head>
<body>
<%=sayHello %>// 使用表达式将对应的字符串输出到页面上
<br>
<%=say()%>// 使用表达式调用对应的方法，输出它的返回值
</body>
</html>
```

3. 代码段

通过使用代码段可以在 JSP 中包含完整的 Java 代码段，其作用是控制 JSP 页面的内部程序流程。代码段的语法格式如下：

```
<% 代码段 %>
```

在 JSP 中编写代码段时，需注意以下三点：

1）代码段并不限于方法、变量和常量的声明；

2）代码段不能像表达式那样直接生成字符串输出；

3）代码段中每条 Java 语句之间用分号分隔。

例如：当使用 if...else... 判断语句时，使用代码段将 if...else... 逻辑判断分隔开，根据 username 具体的数值进行显示，效果如图 2-5 所示。

图 2-5　JSP 代码段实例

实现图 2-5 所示的效果，如示例代码 2-3 所示。

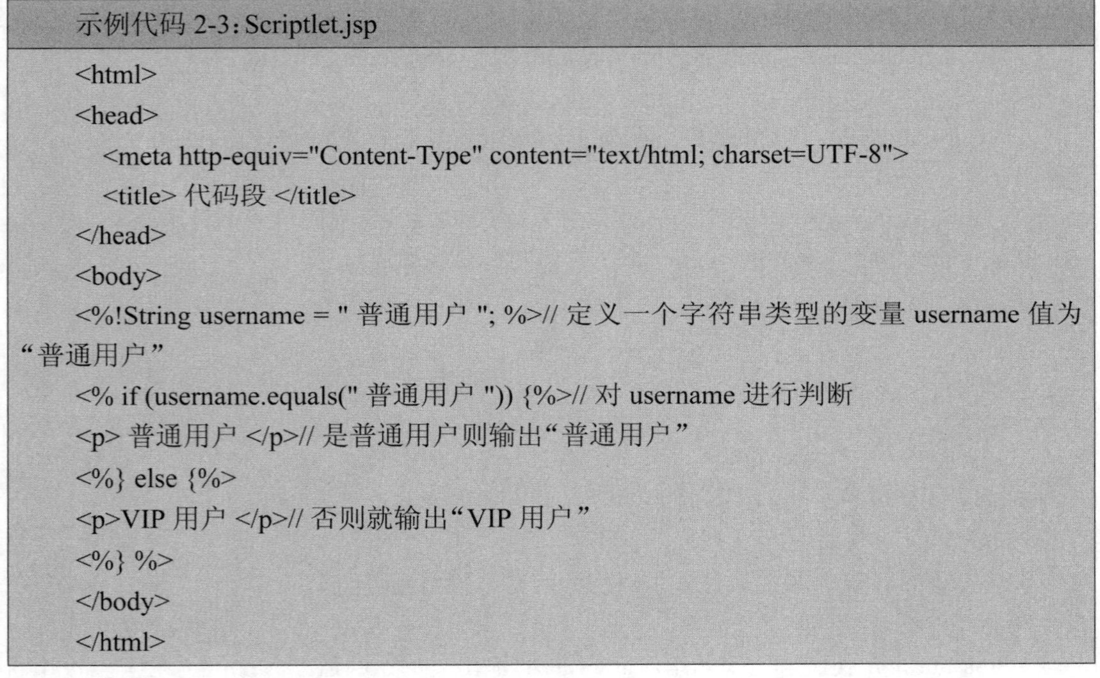

示例代码 2-3：Scriptlet.jsp

```
<html>
<head>
   <meta http-equiv="Content-Type" content="text/html; charset=UTF-8">
   <title> 代码段 </title>
</head>
<body>
<%!String username = " 普通用户 "; %>// 定义一个字符串类型的变量 username 值为
"普通用户"
<% if (username.equals(" 普通用户 ")) {%>// 对 username 进行判断
<p> 普通用户 </p>// 是普通用户则输出"普通用户"
<%} else {%>
<p>VIP 用户 </p>// 否则就输出"VIP 用户"
<%} %>
</body>
</html>
```

4. 注释

注释是在编写程序过程中想要表明代码的含义时进行标注的一种方式，JSP 注释为隐式注释，其注释的内容在客户端浏览时可被隐藏，且在客户端查看 HTML 源代码时不会被看到，所以安全性较高。注释的语法格式如下：

```
<%-- 注释内容 --%>
```

例如：在 JSP 页面上实现获取当前系统的时间并使用注释将“获取当前时间”的中文提示词隐去，效果如图 2-6 所示。

当前时间为：2020-6-8 10:58:07

图 2-6　JSP 注释实例

实现图 2-6 所示的效果，如示例代码 2-4 所示。

示例代码 2-4：javawebdemo.jsp

```
<%@ page language="java" contentType="text/html; charset=GBK"
        pageEncoding="GBK" %>
<html>
<head>
   <meta http-equiv="Content-Type" content="text/html; charset=UTF-8">
   <title> 获取当前时间 </title>
</head>
<body>
<%-- 获取当前时间 --%>
<table>
   <tr>
      <td>
         当前时间为：<%=(new java.util.Date()).toLocaleString() %>
      </td>
   </tr>
</table>
</body>
</html>
```

浏览器查看 HTML 源代码，如图 2-7 所示。

```
<html>
<head>
<meta http-equiv="Content-Type" content="text/html; charset=UTF-8">
<title>获取当前时间</title>
</head>
<body>

<table>
    <tr><td>当前时间为：2020-6-8 10:58:07</td></tr>
</table>

</body>
</html>
```

图 2-7　HTML 源代码

在源代码中，没有看到相对应的“获取当前时间”的中文提示词，表示 JSP 注释属于隐式注释，在源代码中不可见，安全性较高。

技能点三 JSP 指令标记

指令标记主要用于为整体的 JSP 页面提供信息。JSP 指令不会生成任何代码，只是为 JSP 容器提供指导和指示来处理 JSP 页面，其语法格式如下：

```
<%@ 指令名 属性 1= 值 属性 2= 值 %>
```

其中：

1）以“<%@”标记开始，“%>”标记结束；

2）在起始符号“%@”之后和结束符号“%”之前可以加空格，也可以不加；

3）在起始符号中的“<”和“%”之间、“%”和“@”之间以及结束符号中的“%”和“>”之间不能有任何空格。

1.page 指令标记

page 指令标记用于设置当前 JSP 页面的属性，从容器的角度来看，每个 JSP 页面都是一个单独的翻译单元（translation unit）。同一个 Web 应用中的各个 JSP 页面可以有自己的 page 指令，这就能够使各个翻译单元分别有一组不同的指示。

page 指令作用于整个 JSP 页面，一般情况下 page 指令放在整个 JSP 页面的起始位置，一个 JSP 页面中可以使用多个 page 指令，其语法格式如下：

```
<%@page 属性 1= 值 ...%>
```

page 指令有如表 2-1 所示的一些属性，这些属性不一定都要设置，如果未设置，则使用其默认值。

表 2-1 page 指令属性

属性	说明
language	JSP 的脚本语言，目前只能是 Java
extends	容器会将 JSP 翻译为一个 Servlet，该属性用于指定该 Servlet 的基类
import	导入当前 JSP 页面所要用的库类，多个类用逗号分隔。容器把此 JSP 翻译称为 Java 源代码时，这个属性的值会被翻译成多个 import 声明
session	指明当前页面是否创建或维持一个 HTTP 会话
isELIgnored	指明当前页面中的 EL 表达式是否忽略
buffer	指明隐式对象 out 所用的输出缓冲区的大小，默认为 8kB
autoFlush	设置为 true（默认）时，输出缓冲区满则自动进行刷新；设置为 false 时，缓冲区满则抛出异常

续表

属性	说明
errorPage	当前页面出现错误时，将跳转到 errorPage 属性指定的错误处理页面
contentType	告知 JSP 容器当前页面返回给用户时的 HTTP 头部，即 MIME 类型
pageEncoding	指定当前页面的字符格式，默认为 ISO-8859-1
info	通常用于定义 JSP 页面的描述信息。属性值使用 getServletInfo() 方法得到，此方法通常用于获得描述 JSP 文件的信息

例如：使用 page 中的 info 属性显示内容，并设置 page 指令中的 import、buffer 和 errorPage 属性，效果如图 2-8 所示。

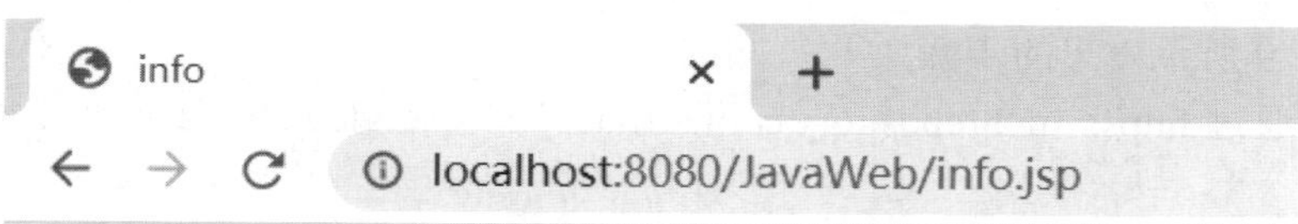

info指令使用

图 2-8　page 指令实例

实现图 2-8 所示的效果，如示例代码 2-5 所示。

示例代码 2-5：info.jsp

```
<%@ page info="info 指令使用 "contentType="text/html;charset=UTF-8"%>
// 设定 info 属性，页面类型为 text/html，字符集为 UTF-8
<%@page import="java.util.*,java.lang.* "%>// 导入 java.util 包和 java.lang 包
<%@page buffer="5kb" autoFlush="false"%>// 设定当前的缓冲区大小为 5kB，并且需要手动刷新缓冲区
<%@page errorPage="error.jsp"%>// 设定当前 JSP 页面出错后跳转到的 JSP 页面为 error.jsp
<html>
  <head><title>info</title></head>
<body>
    <% out.println(getServletInfo()); %>// 使用 getServletInfo 方法获取指令属性 info 中的内容
</body>
</html>
```

2.taglib 指令

taglib 指令的作用是在 JSP 页面中将标签库表述符文件引入该页面中，并设置前缀，利用标签的前缀使用标签库表述文件中的标签，其语法格式如下：

```
<%@taglib uri = " 标签库表述符文件 " prefix = " 前缀名 " %>
```

注意：必须在使用自定义标签之前使用 <%@taglib%> 指令，而且可以在同一个页面中多次使用，但是前缀只能使用一次。

例如：在 taglibdemo.jsp 中使用 taglib 指令标记，引入 jstl 标签库，使用该标签库输出一些中文提示词，效果如图 2-9 所示。

图 2-9 taglib 指令实例

实现图 2-9 所示效果的步骤如下。

第一步：通过 https://mvnrepository.com/artifact/javax.servlet/jstl/1.2 页面下载 jstl 标签库 jar 包。

第二步：将 jar 包粘贴到项目中的 WEB-INF/lib 文件夹中，如图 2-10 所示。

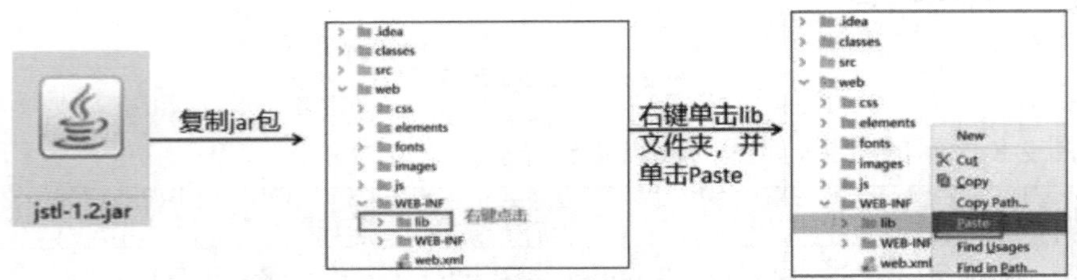

图 2-10 引入 jstl 标签库

具体如示例代码 2-6 所示。

示例代码 2-6：taglibdemo.jsp

```
<%@ page contentType="text/html;charset=UTF-8" language="java" %>
<%@taglib prefix="c" uri="http://java.sun.com/jsp/jstl/core" %>// 使用 taglib 指令引入 jstl 标签库，并定义它的前缀为“c”
<html>
<head><title>taglib 指令标记实例 </title></head>
<body>
<p>
    <c:out value=" 使用 taglib 引入 jstl 标签库 "></c:out>// 使用定义的标签库前缀“c:out”输出 value 中的中文提示词
</p>
</body>
</html>
```

3.include 指令

在设计网站时，有时需要让网站中所有页面的同一位置都包含一段相同的代码，一般情况下是将这部分代码独立出来做成一个文件，然后用 include 指令嵌入每个页面中。这样可以统一修改页面显示，更加高效整洁。

在使用 include 指令过程中，需要注意以下三点：

1）被引入的文件必须遵循 JSP 语法；

2）被引入的文件可以使用任意的扩展名，但都会被 JSP 引擎按照 JSP 页面的处理方式处理；

3）引入和被引入文件中的指令不能冲突（page 指令中的 pageEncoding 和 import 属性除外）。

include 指令的语法格式如下：

```
<%@include file=" 被包含的页面 "%>
```

该指令会使 JSP 页面在编译阶段把其他文件的内容与当前 JSP 合并。

使用了 include 指令的 JSP 页面在转换时，JSP 容器会在其中插入所包含文件的文本或代码，在一些网站中页面的头部和尾部一般都是使用 include 指令引入的，在方便修改的同时也提高了页面的整体性。需要明确的一点是，合并所包含文件的动作是在编译时发生的，而不是在请求时发生的。这说明首先会合并包含文件和被包含文件，然后合并后的整个输出作为一个单元得到编译。如果所包含的文件有所改动，容器是无法知道的，只能重新编译整个单元。

例如：在一个 JSP 页面中使用 include 指令添加一个 index_head 头部文件，效果如图 2-11 所示。

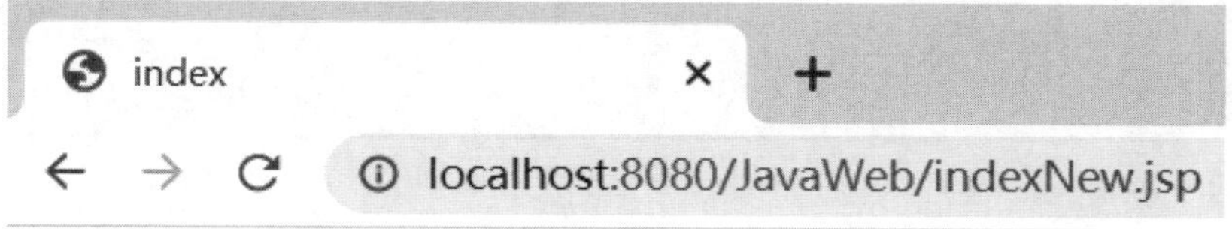

图 2-11　include 指令实例

在 indexNew.jsp 页面中，使用 include 指令标记引入 index_head.jsp 页面，实现图 2-11 所示的效果，如示例代码 2-7 所示。

示例代码 2-7：indexNew.jsp（主页面）

```
<%@ page language="java" contentType="text/html; charset=UTF-8"
    pageEncoding="UTF-8" %>
<html>
```

```
<head>
  <meta http-equiv="Content-Type" content="text/html; charset=UTF-8">
  <title>index</title>
</head>
<body>
<%@include file="index_head.jsp" %>// 引入 index_head 页面
<p> 这是 index 页面中的 body</p>
</body>
</html>
```

index_head.jsp 页面如示例代码 2-8 所示。

示例代码 2-8:index_head.jsp(被引入的头部页面)

```
<%@ page language="java" contentType="text/html; charset=UTF-8"
     pageEncoding="UTF-8"%>
<html>
<head>
  <meta http-equiv="Content-Type" content="text/html; charset=UTF-8">
  <title>index_head 页面 </title>
</head>
<body>
<h1> 这是 index 页面的头部文件 </h1>// 使用 h1 标签并添加一些中文提示词
</body>
</html>
```

技能点四　JSP 动作标记

JSP 动作标记是一种特殊标签,并以前缀 JSP 和其他的 HTML 标签相区别,利用 JSP 动作标记可以实现很多功能,包括动态插入文件、使用 JavaBean 组件、把用户重新定向到另外的页面、为 Java 插件生成 HTML 代码等,其语法格式如下:

```
<prefix:tagName [attribute1=value1]...[attributen=valuen] >
     tagbody
</prefix:tagName>
```

或者

```
<prefix:tagName [attribute1=value1]...[attributen=valuen] / >
```

JSP 动作标记在请求处理阶段按照页面中出现的顺序执行,其与指令标记的不同之处在于，动作标记不使用嵌入的方式来处理该文件，在 JSP 页面运行时才处理文件，执行的速度比较慢，但是可以灵活处理所需文件。

1.param 动作标记

param 动作标记主要用来以“name=value”的形式为其他元素提供附加信息，其基本语法格式和参数如下：

```
<jsp:param name=" 参数名 " value=" 参数值 | <%=expression1%>"/>
```

其中，name 属性表示传递的参数名称；value 属性为该参数的数值，可以使用 JSP 表达式动态引入。

在使用 param 的过程中，需要注意以下几点：

1）属性值必须加上双引号，否则执行时会报错；

2）在 JSP 页面中通过 request.getParameter（“属性名”）来获取参数的值；

3）<jsp:param> 动作标记必须配合 <jsp:include>、<jsp:forward> 或 <jsp:plugin> 等标记使用。

例如：使用 <jsp:param> 动作标记向被包含文件传递参数。通过动作标记 <jsp:param> 引入 number.jsp，实现计算 1 到 300 相加的和，效果如图 2-12 所示。

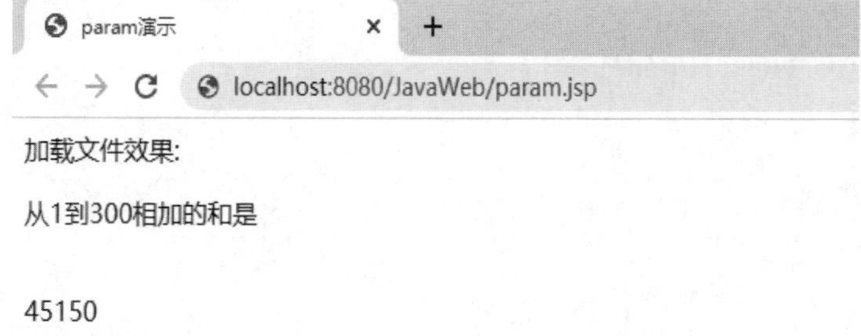

图 2-12　param 动作标记实例

使用 <jsp:include page="number.jsp"> 和 </ jsp:include> 引入对应的页面，具体如示例代码 2-9 所示。

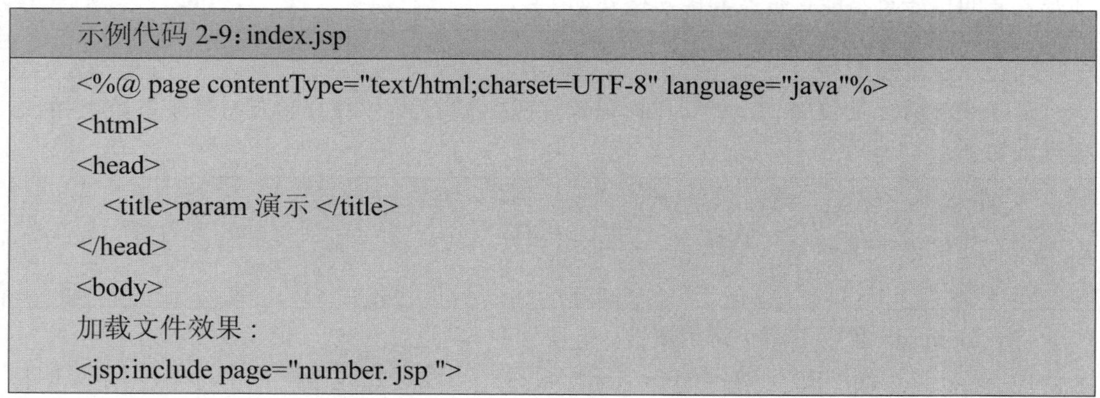

示例代码 2-9：index.jsp

```
<%@ page contentType="text/html;charset=UTF-8" language="java"%>
<html>
<head>
  <title>param 演示 </title>
</head>
<body>
加载文件效果：
<jsp:include page="number. jsp ">
```

```
<jsp:param name="computer" value="300"/>// 使用 param 动作标记传递名为 computer,
数值为 300 的参数至 number.jsp
</jsp:include>
</body>
</html>
```

使用 JSP 脚本通过 request.getParameter（“computer”）获取名称为 computer 属性的值，并将其转换为 int 类型之后利用 for 循环进行求和，并使用 JSP 表达式将总和 sum 输出到页面上，具体如示例代码 2-10 所示。

示例代码 2-10：number.jsp

```
<%@ page contentType="text/html;charset=UTF-8" language="java" %>
<html>
<body>
<%
   String str = request.getParameter("computer");// 获取名为 computer 属性的值
   int n = Integer.parseInt(str);// 将获取的字符串类型数值转化为整型
   int sum = 0;
   for(int i=1;i<=n;i++){// 循环求和
      sum+=i;}
%>
<p> 从 1 到 <%=n%> 相加的和是 </p>
<br>
<%=sum%>
</body>
</html>
```

2.forward 动作标记

forward 动作标记将客户端发出来的请求用于两个页面之间的跳转，它会引起 Web 服务器的请求目标转发。其在服务器进行，不会引起用户端的二次请求，并且在 <jsp:forward> 标签之后的程序将不能执行。其语法格式如下：

```
<jsp:forward page={" 页面 URL"|"<%=expression %>"}/>
```

带参数的情况是使用 param 动作标记放置在标签之中，以便在转发时携带参数，它的语法规则如下：

```
<jsp:forward page=" 页面 URL">
    <jsp:param name=" 属性名 " value=" 属性值 "/>
</jsp:forward>
```

<jsp:forward> 有以下两个常用属性。

1）{page=" 页面 URL"|"<%=expression %>"}：一个表达式或是一个字符串，用于说明将

要定向的文件或者 URL。

2）<jsp:param name=" 属性名 " value=" 属性值 "/>：向一个动态文件发送一个或者多个参数，name 指定参数名称，value 指定参数值。其中，该文件一定是动态文件；如果想传递多个参数，可以在一个 JSP 文件中使用多个 <jsp:param> 标签。

例如：使用 forward 动作标记在页面上显示用户名以及密码，效果如图 2-13 所示。

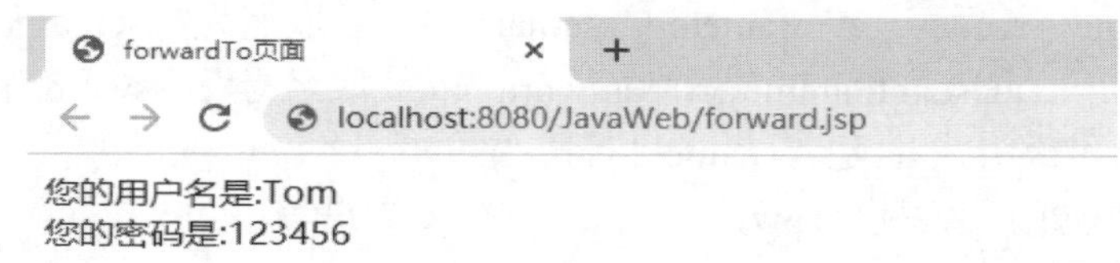

图 2-13　forward 动作标记实例

运行 forward.jsp 之后可以看到跳转之后的页面，通过观察地址栏可以发现它所对应的页面并没有改变，依旧是 forward.jsp，并没有引起客户端的二次请求，而导致地址栏发生变化。

使用带参数的 forward 动作标记定义两个参数 username 和 password 跳转至 forwardTo.jsp 页面。实现图 2-13 所示的效果，如示例代码 2-11 所示。

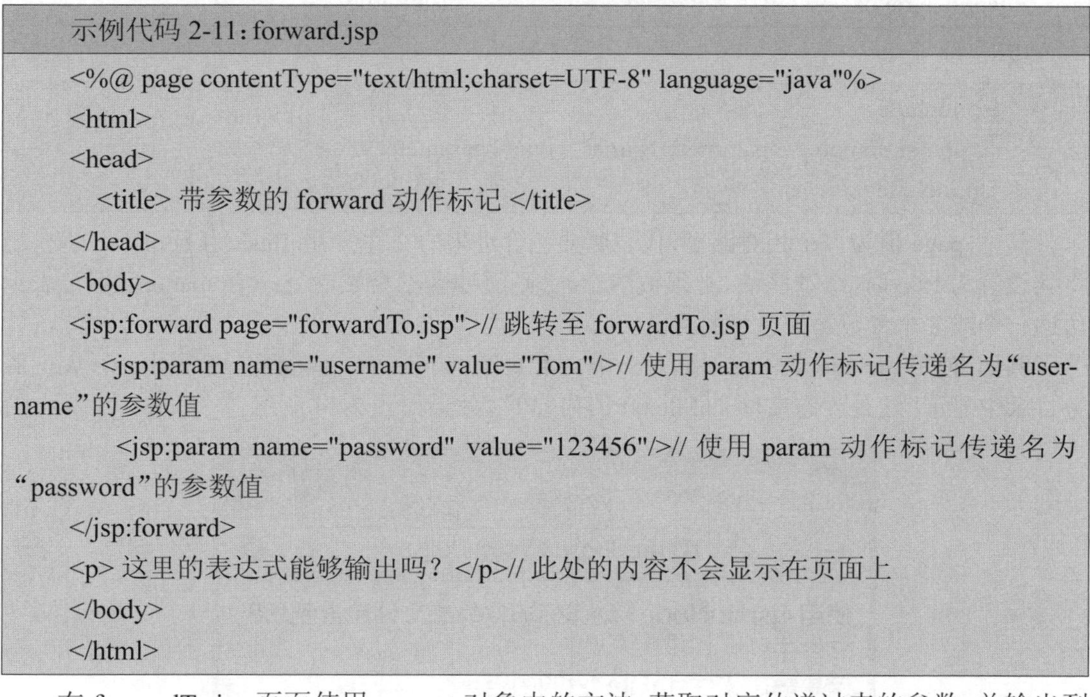

示例代码 2-11：forward.jsp

```
<%@ page contentType="text/html;charset=UTF-8" language="java"%>
<html>
<head>
  <title> 带参数的 forward 动作标记 </title>
</head>
<body>
<jsp:forward page="forwardTo.jsp">// 跳转至 forwardTo.jsp 页面
  <jsp:param name="username" value="Tom"/>// 使用 param 动作标记传递名为“username”的参数值
    <jsp:param name="password" value="123456"/>// 使用 param 动作标记传递名为“password”的参数值
</jsp:forward>
<p> 这里的表达式能够输出吗？ </p>// 此处的内容不会显示在页面上
</body>
</html>
```

在 forwardTo.jsp 页面使用 request 对象中的方法，获取对应传递过来的参数，并输出到页面上，具体如示例代码 2-12 所示。

示例代码 2-12：forwardTo.jsp

```
<%@ page contentType="text/html;charset=UTF-8" language="java" %>
```

```
<html>
   <title>forwardTo 页面 </title>
</head>
<body>
<%
   String name = request.getParameter("username");// 获取名为“username”的属性值
   String pw = request.getParameter("password");// 获取名为“password”的属性值
   out.print(" 您的用户名是 :"+name+"<br/>");
   out.print(" 您的密码是 :"+pw);
%>
</body>
</html>
```

3.include 动作指令

include 动作指令会在当前的 JSP 页面中加入静态和动态的资源。对于静态网页，直接将内容加入 JSP 网页中；对于动态网页则会在编译运行之后再加入 JSP 网页中。其语法格式如下：

```
<jsp:include page="URL 或 <%=expression%>" flush="true"/>
```

或者

```
<jsp:include>
   {<jsp:param name="parameterName" value="parameterValue"/>}*
</jsp:include>
```

其中，page 值为一个相对路径，代表所要包含进来的文件位置；flush 为 boolean 类型，当将其设置为 true 时，缓冲区满，必须被清空，否则缓冲区不会被清空；<jsp:param> 标记表示传递一个或多个参数给被包含的网页。

例如：使用 include 动作标记插入一个文本文件 staFile.txt，并在当前页面同一个 Web 服务目录中显示“这是静态文件 staFile.txt 的内容 !”，效果如图 2-14 所示。

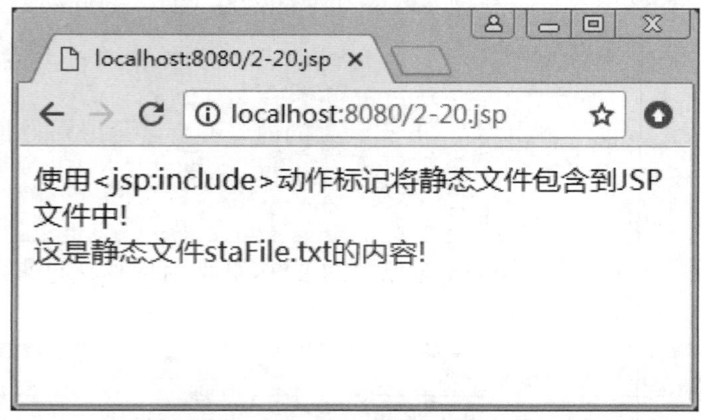

图 2-14　include 动作标记实例

实现图 2-14 所示的效果，具体步骤如下。

第一步：新建一个名为 staFile.txt 的文本文件，设置中文提示词的样式，如示例代码 2-13 所示。

示例代码 2-13：staFile.txt

```
<font color="blue" size="3">// 设置字体大小以及颜色
<br> 这是静态文件 staFile.txt 的内容！
</font>
```

第二步：创建 index.jsp 页面，使用 include 动作标记将 staFile.txt 添加到 index 页面中，如示例代码 2-14 所示。

示例代码 2-14：index.jsp

```
<%@ page contentType="text/html;charset=UTF-8" %>
<html>
<body>
    使用 &lt;jsp:include&gt; 动作标记将静态文件包含到 JSP 文件中 !// 利用 &lt; 和 &gt; 来转义“<”和“>”
</hr>
    <jsp:include page="staFile.txt"  />
</body>
</html>
```

include 指令标记与 include 动作标记功能相似，但也有区别，具体如表 2-2 所示。

表 2-2　include 指令标记与 include 动作标记的区别

区别	include 指令标记	include 动作标记
语法格式	<%@include file="..."%>	<jsp:include page="...">
作用时间	页面转换期间	请求期间
是否影响主页面	影响	不影响
内容变化时是否需要手动修改包含页面	需要	不需要
编译速度	较慢	较快
执行速度	较快	较慢
灵活性	较差	较好
对被包含文件的约定	有约定（不能重复 html 标签，如 <html></html>）	无约定

运用 JSP 基础知识以及 JSP 指令完成网上书城项目实战中登录页面的创建，效果如图 2-15 所示。

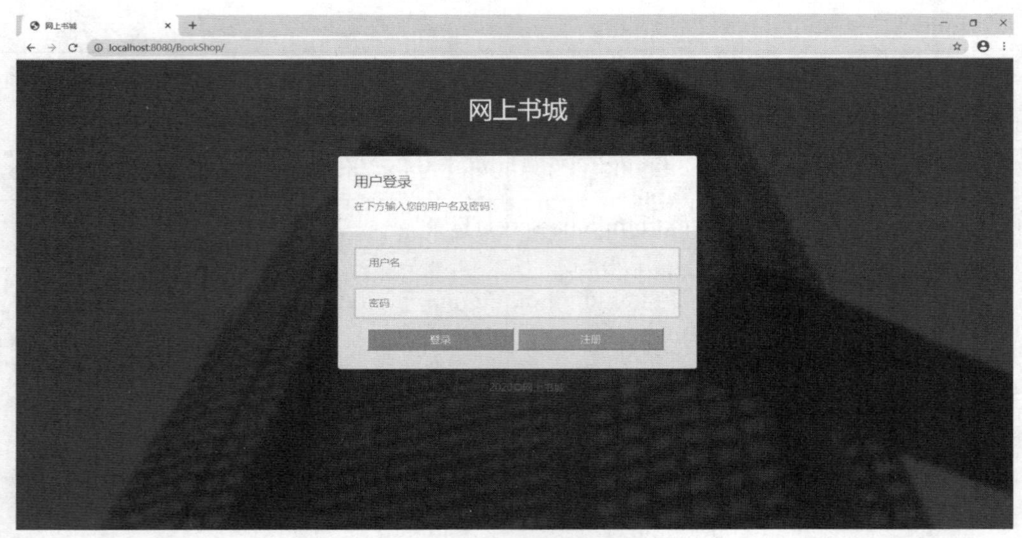

图 2-15　网上书城登录页面

具体步骤如下。

第一步：创建 login.jsp 页面。主要的 JSP 页面创建位置为 web 文件夹，鼠标右键单击 web 文件夹新建 JSP 页面，并命名为 login，如图 2-16 所示。

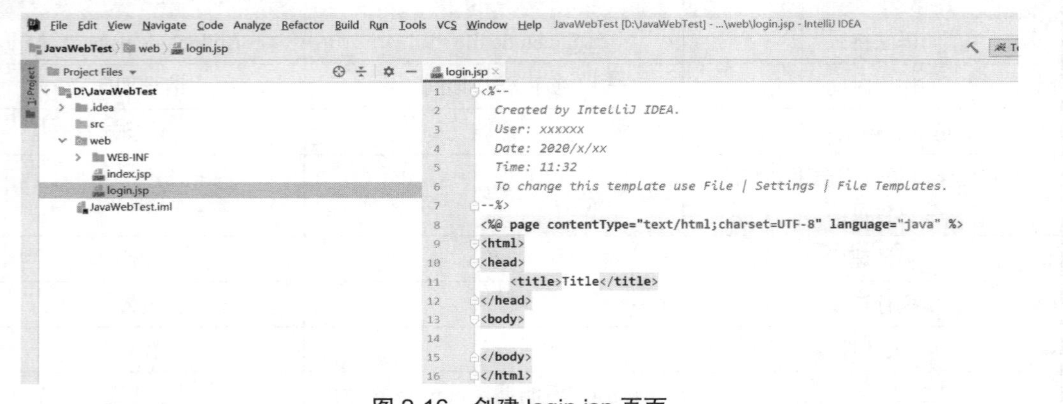

图 2-16　创建 login.jsp 页面

第二步：构建静态资源的目录。在 web 文件夹目录下创建 css 文件夹放置 CSS 样式文件，fonts 文件夹存放文字样式内容，images 文件夹存放网站图片资源内容，js 文件夹存放 JavaScript 文件，elements 文件夹存放网站头部文件、搜索框和底部文件。具体目录如图

2-17 所示。

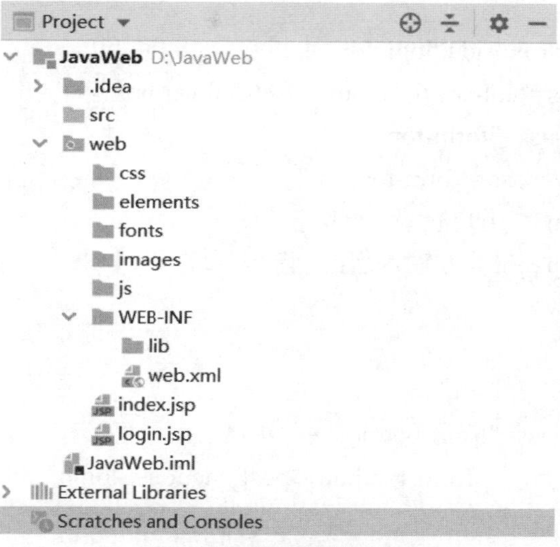

图 2-17　静态资源目录

第三步：编写 login.jsp 页面，验证用户登录信息，并引入头部和底部文件，如示例代码 2-15 所示。

示例代码 2-15：login.jsp

```
<%@ page language="java" import="java.util.*"
      contentType="text/html; charset=UTF-8"%>
<jsp:include page="elements/index_head.jsp"></jsp:include>
<%// 查询 session 数据，用户是否已经登录，若已经登录则直接跳转至主页
   String username = (String) session.getAttribute("loginuser");
   if (username != null)
      response.sendRedirect("books");
%>
<link rel="stylesheet" href="css/bootstrap1.min.css">// 引入相关样式
<link rel="stylesheet" href="css/font-awesome.min.css">
<link rel="stylesheet" href="css/form-elements.css">
<link rel="stylesheet" href="css/style2.css">
<body>
<div class="top-content">
   <div class="inner-bg">
   <div class="container">
         <div class="row book">
            <div class="col-sm-8 col-sm-offset-2 text">
                  <h1><strong> 网上书城 </strong></h1>
```

```
            </div>
        </div>
        <div class="row" id="login">
            <div class="col-sm-6 col-sm-offset-3 form-box">
                <div class="form-top">
                    <div class="form-top-left">
                        <h3> 用户登录 </h3>
                        <p> 在下方输入您的用户名及密码：</p>
                    </div>

                </div>
                <div class="form-bottom">
                            <form method="post" action="login" onsubmit="return check()"
class="login-form">
                        <div class="form-group">
                            <label class="sr-only" for="username"> 用户名：</label>

                            <input type="text" name="username" placeholder=" 用户名 " class="-
form-username form-control" id="username" onblur="isUsernameNull()" ><span id="user-
null"></span>
                        </div>
                        <div class="form-group">
                                <label class="sr-only" for="password"> 密码：</label>
                                <input type="password" name="password" placeholder=" 密码 "
class="form-password form-control" id="password" onblur="isPasswordNull()"><span id="p-
wdnull"></span>
                        </div>
                        <button class="input-btn" type="submit" name="submit" value=" 登录 "
style="margin-left: 4%; width: 45%;background:#ff5722b8; color: white"> 登录 </button>
                        <button class="input-reg" type="button" name="register" value=" 注册 "
onclick="window.location='register.jsp';" style="width:45%;background:#19c880; color: white"
> 注册 </button>
                    </form>
                </div>
            </div>
        </div>

    </div>
```

```
</div>
<script src="js/jquery-1.11.1.js"></script>
    <script src="js/bootstrap.min.js"></script>
<script src="js/jquery.backstretch.js"></script>
<script src="js/scripts.js"></script>
</body>
<jsp:include page="elements/main_bottom.html"/>
```

第四步：编写用户名和密码内容检查的 JavaScript 部分，位于 login.jsp 页面中，如示例代码 2-16 所示。

示例代码 2-16：login.jsp

```
<script type="text/javascript">
  // 非空检查
  function isUsernameNull(){
    var username = document.getElementById("username").value;
    var usernull = document.getElementById("usernull");
    if(username == null || username == ""){
      usernull.innerHTML = "<font color=\"red\"> 用户名不能为空！ </font>"
      return false;
    }else       usernull.innerHTML = ""
    return true;
  }
  function isPasswordNull(){
    var password = document.getElementById("password").value;
    var pwdnull = document.getElementById("pwdnull");

if(password == null || password == ""){
      pwdnull.innerHTML = "<font color=\"red\"> 密码不能为空！ </font>"
      return false;
    }else
      pwdnull.innerHTML = ""
    return true;
  }
  function check(){
    if(!isUsernameNull()){
      return false;
    }else if(!isPasswordNull()){
      return false;
```

```
        }
        return true;
    }
</script>
```

第五步：编写 include 动作标记引入的头部文件内容，文件目录位于 elements 文件夹下，如示例代码 2-17 所示。

示例代码 2-17：index_head.jsp

```
<%@ page language="java" import="java.util.*" pageEncoding="UTF-8"%>
<!DOCTYPE html PUBLIC "-//W3C//DTD XHTML 1.0 Transitional//EN" "http://www.w3.org/TR/xhtml1/DTD/xhtml1-transitional.dtd">
<html xmlns="http://www.w3.org/1999/xhtml">
    <head>
      <meta http-equiv="pragma" content="no-cache">
      <meta http-equiv="cache-control" content="no-cache">
      <meta http-equiv="expires" content="0">
      <meta http-equiv="Content-Type" content="text/html; charset=UTF-8" />
      <title> 网上书城 </title>
      <link type="text/css" rel="stylesheet" href="css/style.css" />//logo 在 css 文件中进行引入
    </head>
```

第六步：编写 include 动作标记引入的底部文件内容，文件目录位于 elements 文件夹下，如示例代码 2-18 所示。

示例代码 2-18：main_bottom. html

```
<html>
<div class="footer">
    <div class="container">
        <div class="footer-main">
            <div class="container">
                <div class="copy-rights-main">
                    <p>2020&copy; 网上书城 </p>
                </div>
            </div>
        </div>
    </div>
</div>
</html>
```

完成上述操作，网上书城项目的登录页面就已经编写好，效果如图 2-15 所示。

本次任务通过对 JSP 基本语法、指令标记、动作标记的介绍和探析，使读者重点熟悉 JSP 中常用的语法和指令及引入外部文件的方法，学习 JSP 页面的基本结构和布局，了解 include 指令标记和动作标记之间的区别及它们的优缺点，为他们以后制作整体项目打下良好的基础。

final	最终的	page	页
include	包括	forward	前进
declaration	声明	element	元素

1. 选择题

1）要运行 JSP 程序，下列说法不正确的是（　　）。

A. 服务器端需要安装 Servlet 容器，如 Tomcat 等

B. 客户端需要安装 Servlet 容器，如 Tomcat 等

C. 服务器端需要安装 JDK

D. 客户端需要安装浏览器，如 IE 等

2）下面关于 JSP 注释的说法，正确的是（　　）。

A.JSP 注释语法格式为 <!-- 注释信息 -->　　B.JSP 注释不会被发送到客户端

C.JSP 注释会在服务器端编译　　D.JSP 注释与 HTML 注释是一样的

3）在 JSP 页面中有如下代码：

```
<%@page import="java.util.Date"%>
<%@page import="java.io.*"%>
```

下面选项中，与之功能相同的是（　　）。

A.<%@page import="java.util.Date java.io.*"%>

B.<%@page import="java.util.Date;java.io.*"%>

C.<%@page import="java.util.Date,java.io.*"%>

D.<%@page import="java.util.Date!java.io.*"%>

4）在 JSP 中导入自定义标签的方式正确的是（　　）。

A.web.xml 文件的 <taglib>
B.<%@ taglib uri="" prefix=""%>
C.<jsp:taglib uri="" prefix=""%>
D. 无须导入，可直接使用

5）能够获取当前页面信息并调用页面方法的对象是（　　）。

A.request　　B. page　　C. PageContext　　D. session

2. 简答题

include 指令标记和 include 动作标记的区别有哪些？

第三章　网上书城项目主页设计

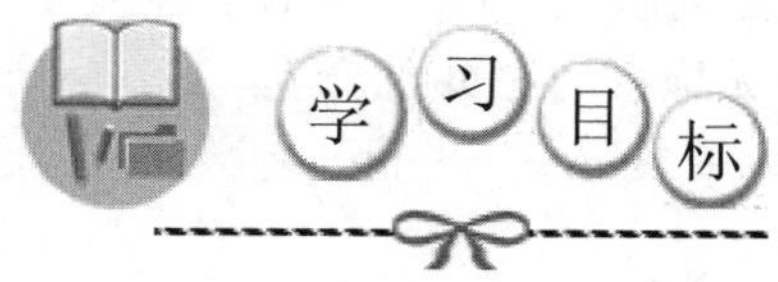

通过学习 JSP 内置对象知识，结合 JSP 基本语法以及 HTML 基础知识，了解和掌握 JSP 内置对象、EL 表达式和 JSTL 的基本知识，具有运用 JSP 内置对象、EL 表达式和 JSTL 的相关知识和编写网上书城项目主页的能力。在任务实现过程中：

- 了解 JSP 内置对象的基本概念；
- 掌握 JSP 内置对象的基本方法；
- 掌握 Cookie 对象的概念和基本方法；
- 掌握 EL 表达式和 JSTL 的使用方法。

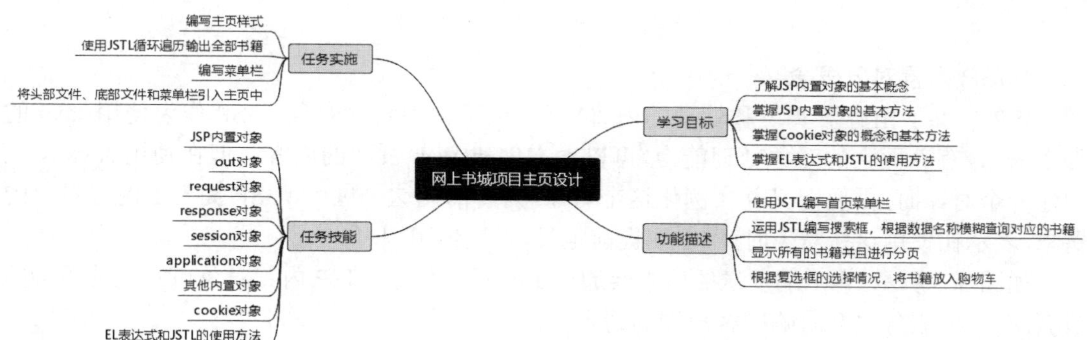

【情境导入】

数据的交互以及传递是网页的特点，对于页面与页面之间的数据交互和页面与后台的数据交互，只运用 JSP 基础知识以及 JSP 指令不足以满足这些需求，所以本次的任务主要是运用 JSP 内置对象、EL 表达式和 JSTL 标签库来实现网上书城项目的主页面设计。

【功能描述】

- 使用 JSTL 编写首页菜单栏。
- 运用 JSTL 编写搜索框，根据数据名称模糊查询对应的书籍。
- 显示所有的书籍并进行分页。
- 根据复选框的选择情况，将书籍放入购物车。

技能点一　JSP 内置对象

1.JSP 内置对象概念

JSP 提供了由容器实现和管理的内置对象，也可称为隐含对象。JSP 作为使用 Java 的脚本语言，有着巨大的对象处理能力，可以动态创建 Web 页面的内容。但在使用 Java 语法创建一个对象时，需要定义并实例化这个对象，为了简化这一过程，JSP 提出了内置对象的概念，表示在使用这些对象时，不需要提前定义，直接使用即可。

如图 3-1 所示，客户端发送的请求会通过 JSP 页面传递，JSP 内置对象可实现前后端的数据交互，使服务器及时响应客户端的请求。

图 3-1　JSP 内置对象交互前后端数据的过程

2. 内置对象

在 JSP 中预先定义了 9 个内置对象，包含输入 / 输出对象、作用域通信对象、Servlet 对象和错误对象，分别为 out、request、response、session、application、config 、page、pageContext 和 exception，如图 3-2 所示。

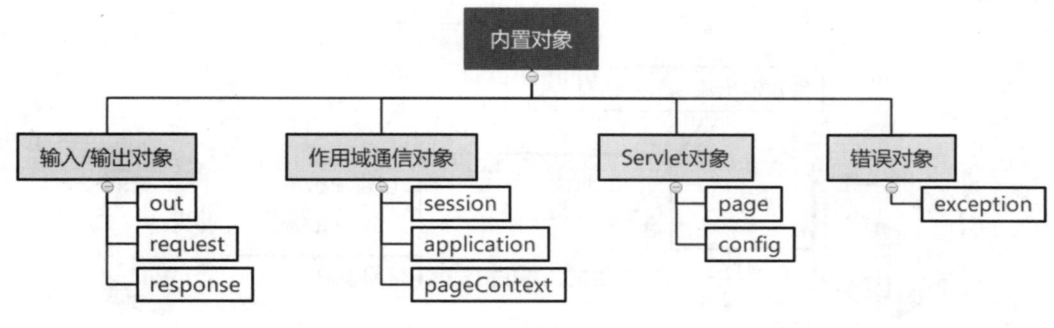

图 3-2　九大内置对象

每个内置对象的作用是不一样的，内置对象的基本作用如表 3-1 所示。

表 3-1　内置对象的基本作用

名称	说明
out	实际上是使用 PrintWrite 类来向客户端浏览器输出数据
request	代表客户端的请求，通过它可以获得客户端提交的数据，如表单中的数据网页地址后带有参数等
response	代表客户端向服务器的响应
session	保持在服务器与一个客户端之间需要保留的数据，当客户关闭网站的所有网页时，session 变量会自动清除
applica-tion	可以用来提供一些全局数据、对象，在 Web 服务器开始提供 Web 服务时就会被创建，一直保持到服务器关闭为止
config	用于取得服务器的配置信息
page	代表 JSP 页面对应的 Servlet 类实例
pageCon-text	取得任何范围的参数，通过它可以获取 JSP 页面的 out、request、reponse、session、application 等对象
exception	含有只能由指定的 JSP“错误处理页面”访问的异常数据

技能点二　out 对象

1.out 对象概述

out 对象作为 JSP 编程过程中常用的对象之一，被封装为 Javax.servlet.jsp.jspWriter 接口，主要是向 Web 浏览器输出各种数据类型的内容，并且管理应用服务器上的输出缓冲区。若 out 对象输出包含 HTML 标记元素，这些标记将会被浏览器正确解释，等效于直接在设计页面时输入的 HTML 文本。

out 对象的工作流程如图 3-3 所示，由图可知 out 对象响应客户端请求，创建输出流以显示信息。

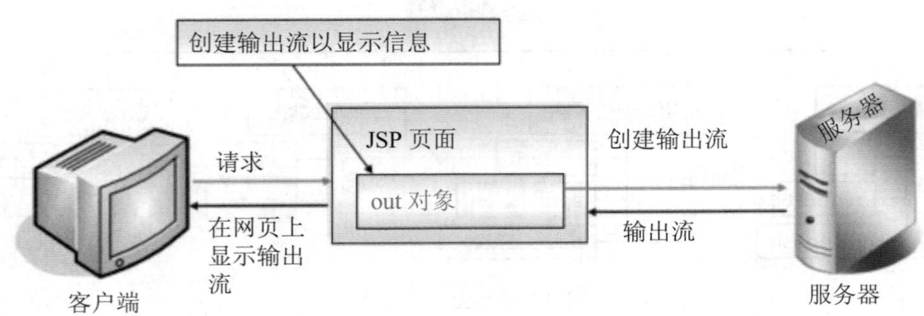

图 3-3 out 对象工作流程

在使用 out 对象的过程中，一般会用到输入输出以清除缓冲相关的方法， out 对象常用的方法如表 3-2 所示。

表 3-2 out 对象常用方法

方法	说明
print()	输出数据，输出完毕后，并不结束该行
println()	输出数据，在输出完毕后，会结束当前行，下一个输出语句将在下一行开始输出
newLine()	输出一个换行符号
clearBuffer()	清除缓冲区里的数据，并且把数据写到客户端去
clear()	清除缓冲区里的数据，但不把数据写到客户端去
getRemaining()	获取缓冲区中没有被占用的空间的大小（kB）
flush()	输出缓冲区里的数据。该方法先将之前缓冲区中的数据输出至客户端，然后再清除缓冲区中的数据
getBufferSize()	获取当前缓冲区的大小（kB），可以通过 page 指令来调整缓冲区的大小
isAutoFlush()	判断缓冲区是否自动刷新
close()	关闭输出流

2.out 对象实例

【实例】应用 out 对象输出缓冲区。

使用 out 对象可实现在 Web 浏览器输出各种数据类型的内容，并且管理应用服务器上的输出缓冲区。使用 out 对象练习输出缓冲区相关方法，实现效果如图 3-4 所示。

图 3-4　out 对象实例

实现图 3-4 所示效果，如示例代码 3-1 所示。

示例代码 3-1：out.jsp

```
<%@ page language="Java" contentType="text/html; charset=UTF-8"
      pageEncoding="UTF-8" %>
<html>
<head>
  <meta http-equiv="Content-Type" content="text/html; charset=UTF-8">
  <title>out 对象 </title>
</head>
<body>
以下是 out 对象其他常用方法的使用：
<hr>
获取缓存大小：<%=out.getBufferSize() %>
<br>
获取剩余缓存大小：<%=out.getRemaining() %>
<br>
判断是否自动刷新：<%=out.isAutoFlush() %>
<br>
<%
  out.print(" 知识改变命运，技术改变生活！ <br>");
  out.print(" 当前可用缓冲区大小：" + out.getRemaining() + "<br>");
  out.flush();// 输出缓冲区中的数据
  out.print(" 当前可用缓冲区空间大小：" + out.getRemaining() + "<br>");
 out.close();// 关闭输出流，在这之后的输出均不会显示在页面上
  out.print(" 这一行代码不会被输出在页面上 ");
%>
```

```
    </body>
</html>
```

技能点三　request 对象

1.request 对象概述

request 对象被用于封装用户提交的信息数据，是 HttpServletRequset 类的一个对象。其工作流程为当服务器接收到客户端发出的 HTTP 请求后，将创建 request 对象，并解析请求数据，且将其保存在 Web 容器中（内存中），当服务器响应请求并将响应数据发送到客户端后，Web 容器自动清除 request 对象，并释放内存，如图 3-5 所示。

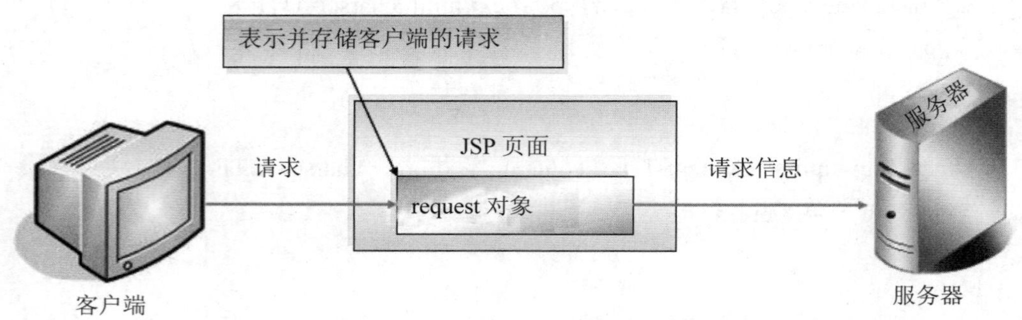

图 3-5　request 对象工作流程

在使用 request 对象的过程中，会用到获取属性内容的方法，request 对象常用的方法如表 3-3 所示。

表 3-3　request 对象常用方法

方法	说明
getAttribute(String name)	返回 name 指定的属性值，若不存在指定的属性，就返回 null
setAttribute(String name，java.lang.Object obj)	设置名为 name 的 request 参数值为 obj
getAttributes()	返回 request 对象的所有属性的名字集合，结果集是一个枚举类（Enumeration）的实例
getCookies()	返回客户端的所有 Cookie 对象，结果是一个 Cookie 数组
getCharacterEncoding()	返回请求中的字符编码方式

续表

方法	说明
getContentLength()	以字节为单位返回客户端请求的大小。如果无法得到该请求的大小，则返回 -1
getProtocol()	返回通信协议
getScheme()	返回请求的方式
getServerName()	返回服务器名称
getServerPort()	返回通信的端口号
getRemoteAddr()	返回使用者的 IP 地址
getRemoteHost()	返回主机地址
getHeader(String name)	用于获得 HTTP 协议定义的文件头信息
getHeaders(String name)	用于返回指定名字的 request Header 的所有值，其结果是一个枚举类（Enumeration）的实例

拓展：HTTP 请求报文和 GET、POST 请求方法。

1.HTTP 请求报文

HTTP 请求报文结构包含请求行、请求头、空行和请求体 4 部分，具体结构如图 3-6 所示。

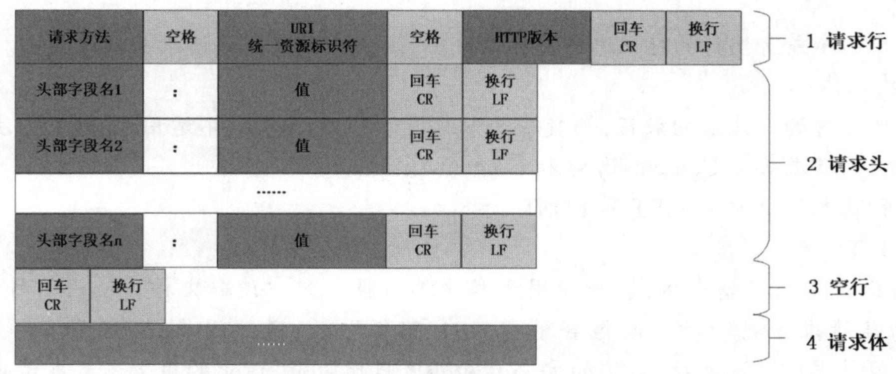

图 3-6　HTTP 请求报文结构

HTTP 协议在使用 TCP 传输过程中需要经过三次握手，在三次握手之后，服务器会接收到客户端发送的一个请求报文，请求报文的格式如图 3-7 所示。

```
▼Request Headers
  :authority: securepubads.g.doubleclick.net
  :method: GET
  :path: /gpt/pubads_impl_rendering_2019121002.js
  :scheme: https
  accept: */*
  accept-encoding: gzip, deflate, sdch
  accept-language: zh-CN,zh;q=0.8
  cache-control: no-cache
  cookie: id=2226fee7c9c000cc||t=1574039534|et=730|cs=002213fd48f2bb4c21410012f9
  pragma: no-cache
  user-agent: Mozilla/5.0 (Windows NT 10.0; WOW64) AppleWebKit/537.36 (KHTML, like Gecko) Chrom
  e/49.0.2623.112 Safari/537.36
```

图 3-7 请求报文格式

(1)请求行

在图 3-7 中,请求行的内容为前三行的内容,包含请求的方法、URI 和 HTTP 协议三个字段,每行的内容用空格进行分割,每个字段的内容用回车或者换行符结尾的方式进行分割。从图 3-7 中可以看出请求方法使用的是 GET 方法,除了 GET 方法,请求的方法还有 POST、HEAD、PUT、DELETE、OPTIONS、TRACE、CONNECT,其中 GET、POST 最为常用。

(2)请求头

请求头用来描述服务器的基本信息,其目的是通过获取这些描述服务器的数据信息,从而通知客户端如何处理它回送的数据。请求头由键 / 值对组成,每行代表一对,键和值的信息用冒号进行分割。

(3)空行

空行用回车或者换行的标识进行内容的分割,用来告诉服务器请求头到此为止。

(4)请求体

请求体包含的是请求的数据,一般情况下适用于 POST 方法,不适用于 GET 方法,和请求数据相关的请求头是 Content-Type 和 Content-Length。

2. 两种基本请求方法:GET 和 POST

(1)GET

GET 是最常用的请求方式,一般用于客户端从服务器中读取文档。使用该种方式需要注意的是请求参数和对应的值会跟在 URL 路径之后,通过问号("?")和 and 连接符("&")以及等号("=")连接,数据的安全性和保密性比较低,请求的报文不存在请求体。

(2)POST

POST 和 GET 请求方式一样,是一种常用的请求方式, POST 因其可以将传输的数据封装在报文的请求中,对传输大小没有限制,能够弥补 GET 方法的不足,安全性和保密性比较高。

2. request 对象实例

【实例】实现使用 request 对象获取表单数据。

应用 request 对象的 getParameter 方法获取 HTML 页面中文本框元素和按钮元素所提交的信息。表单输入效果如图 3-8(a)所示,单击"submit"按钮传递数值之后的效果如图 3-8(b)所示。

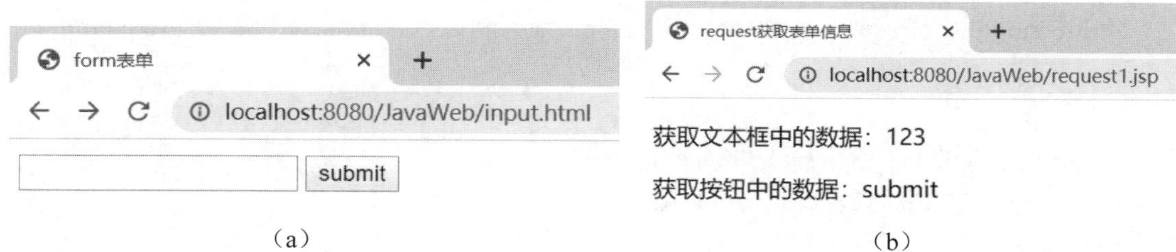

图 3-8　request 对象实例

编写 input.html 页面，使用 form 表单，指定 request1.jsp 页面处理此表单数据，传递 username 以及 submit 按钮信息，具体如示例代码 3-2 所示。

示例代码 3-2：input.html

```
<html>
<head>
<meta charset="UTF-8">
<title>form 表单 </title>
</head>
<body>
<form action="request1.jsp" method="post" >
    <input name="username" type="text" >
    <input name="submit" type="submit" value="submit">
</form>
</body>
</html>
```

编写 request1.jsp 页面，通过 page 配置页面信息，使用 request.getParameter 方法获取 input.html 页面中文本框所输入的数据，使用 request.getParameter 方法获取 input.html 页面中按钮的数据，并通过 JSP 表达式输出到页面上，具体如示例代码 3-3 所示。

示例代码 3-3：request1.jsp

```
<%@ page language="Java" contentType="text/html; charset=UTF-8"
      pageEncoding="UTF-8" %>
<html>
<head>
   <meta http-equiv="Content-Type" content="text/html; charset=UTF-8">
   <title>request 获取表单信息 </title>
</head>
<body>
<%
    String name = request.getParameter("username");// 使用 getParameter 方法获取 username 属性值
```

```
    String but = request.getParameter("submit");// 使用 getParameter 方法获取 submit 属性值
%>
<p> 获取文本框中的数据：<%=name %></p>
<p> 获取按钮中的数据：<%=but %></p>
</body>
</html>
```

【实例】实现使用 request 对象处理中文乱码。

在使用 request 对象接收客户端提交的汉字时，会出现乱码问题。例如在上一个实例中，若将 submit 修改为“按钮”，在进行提交时就会出现乱码，如图 3-9（a）所示。在这种情况下，就要进行特殊的处理。将获取的信息使用 ISO-8859-1 编码形式存储到字节数组 c 中，之后再将这个数组转换为字符串对象。输入“123”之后的运行效果如图 3-9（b）所示。

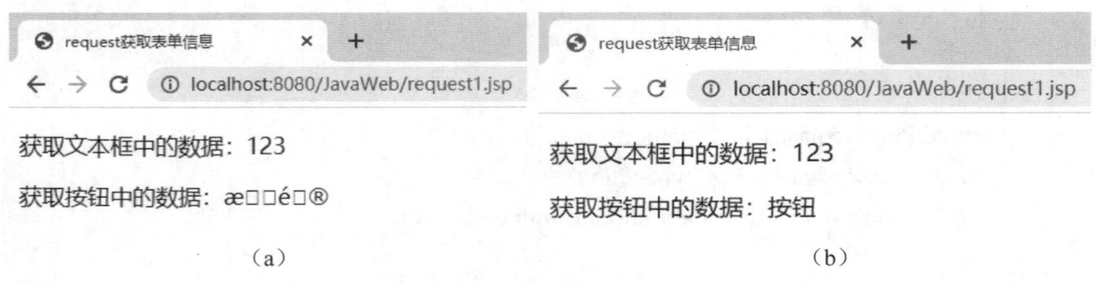

（a） （b）

图 3-9 request 对象处理中文乱码实例

修改 request1.jsp 页面，如示例代码 3-4 所示。

示例代码 3-4：request1.jsp

```
<%@ page language="Java" contentType="text/html; charset=UTF-8"%>
<html>
<head>
<title>request 获取表单信息 </title>
</head>
<body>
<%
    String name = request.getParameter("username");
    String but = request.getParameter("submit");
    byte c[] = name.getBytes("ISO-8859-1");
    name = new String(c);

%>
```

```
<p> 获取文本框中的数据：<%=name %></p>
<p> 获取按钮中的数据：<%=but %></p>
</body>
</html>
```

【实例】实现使用 request 对象进行跳转。

应用 request 对象的跳转方法，一般在 Servlet 跳转到指定页面时使用，特点是只经过一次请求，发出跳转请求的页面和跳转到的页面均可以共享 request 中的数据。例如跳转至 successTest.jsp 页面，它的语法格式如下：

```
request.getRequestDispatcher("successTest.jsp").forward(request, response);
```

例如：使用 request 对象 forward 跳转方法，从 Resquestforward.jsp 页面跳转至 testForward.jsp 页面，实现共享 request 数据并进行页面的跳转，效果如图 3-10 所示。

图 3-10　request 对象跳转实例

在图 3-10 中可以看到访问 Requestforward.jsp 页面后，地址栏应显示为 testForward.jsp 页面，但地址栏并没有发生改变，还是开始访问的 Requestforward.jsp 页面。

在 testForward.jsp 页面中设置名为“test”的 request 参数值为“hello”，并使用 forward 方法进行跳转，具体如示例代码 3-5 所示。

示例代码 3-5：Resquestforward.jsp

```
<%@ page language="java" contentType="text/html; charset=UTF-8"
    pageEncoding="UTF-8"%>
<html>
<head>
  <meta http-equiv="Content-Type" content="text/html; charset=UTF-8">
  <title>forward 转发 </title>
</head>
<body>
<%
  request.setAttribute("test","hello");// 设置名为 test 的属性值为 hello
```

```
    request.getRequestDispatcher("testForward.jsp").forward(request, response);// 跳转至
testForward.jsp 并携带 request 所设置的属性值
%>
</body>
</html>
```

编写 testForward.jsp，通过 getAttribute（“test”）方法获取对应的属性值，并将其输出到页面上，具体如示例代码 3-6 所示。

示例代码 3-6：testForward.jsp

```
<%@ page language="java" contentType="text/html; charset=UTF-8"
    pageEncoding="UTF-8"%>
<html>
<head>
  <meta http-equiv="Content-Type" content="text/html; charset=UTF-8">
  <title> 获取 forward 共享数据页面 </title>
</head>
<body>
<%
  String val = (String)request.getAttribute("test");// 获取对应的属性值
%>
<p> 获取 Resquestforward.jsp 转发传递过来的 test 数值的值：<%=val %></p>// 显示
在页面上
</body>
</html>
```

技能点四　response 对象

1.response 对象概述

response 对象用于动态响应客户端请求，控制发送给用户的信息，并生成动态响应。response 实现 Javax.servlet.http.HttpServletResponse 接口，是 HttpServletResponse 的实例，封装了 JSP 产生的响应客户端请求的有关信息。response 对象工作流程如图 3-11 所示。

如图 3-11 所示，服务端响应客户端的需求，将处理结果以 response 对象的方式返回给 JSP 引擎，JSP 引擎和 Web 服务器根据 response 对象最终生成 JSP 页面，返回给客户端浏览器。

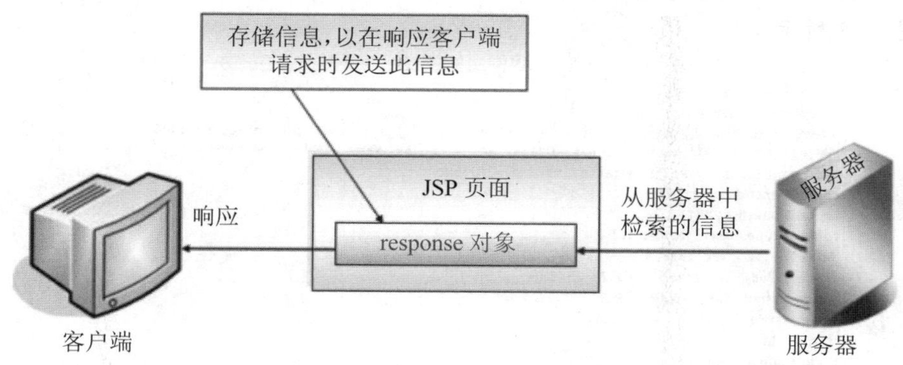

图 3-11　response 对象工作流程

在使用 response 对象的过程中，会用到回应请求和设置返回值等常用方法，如表 3-4 所示。

表 3-4　response 对象常用方法

方法	说明
setContentType(String type)	动态响应 contentType
encodeURL()	使用 sessionId 来封装 URL
flushBuffer()	强制将当前缓冲区的内容发送到客户端
getBufferSize()	返回缓冲区的大小
addHeader(String name,String value)	添加 HTTP 头文件，该 header 将会传递到客户端
setHeader(String name,String value)	设置 HTTP 应答报文的头部字段和值以及自动更新
sendRedirect(String redirectURL)	将客户端重定向到指定 URL
SetStatus(int n)	设置 HTTP 返回值的状态值
addCookie(Cookie cookie)	添加一个 Cookie 对象

拓展：HTTP 响应报文。

HTTP 响应报文包含状态行、响应头、空行和响应体 4 部分，具体结构如图 3-12 所示。

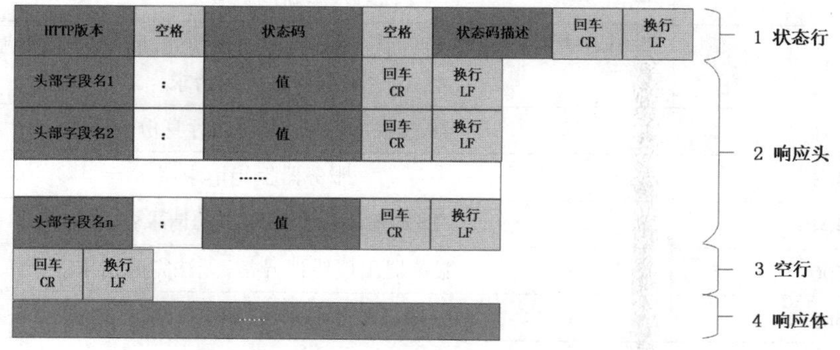

图 3-12　HTTP 响应报文

当收到 GET 或 POST 等方法发来的请求后，服务器就要对报文进行响应。响应报文的格式如图 3-13 所示。

```
▼Response Headers
    accept-ranges: bytes
    alt-svc: quic=":443"; ma=2592000; v="46,43",h3-Q050=":443"; ma=2592000,h3-Q049=":443"; ma=2592
    000,h3-Q048=":443"; ma=2592000,h3-Q046=":443"; ma=2592000,h3-Q043=":443"; ma=2592000
    cache-control: private, immutable, max-age=31536000
    content-encoding: gzip
    content-length: 24811
    content-type: text/javascript
    date: Tue, 24 Dec 2019 03:38:29 GMT
    expires: Tue, 24 Dec 2019 03:38:29 GMT
    last-modified: Tue, 10 Dec 2019 17:29:18 GMT
    server: sffe
    status: 200
    timing-allow-origin: *
    vary: Accept-Encoding
    x-content-type-options: nosniff
    x-xss-protection: 0
```

图 3-13 响应报文格式

(1)状态行

状态行一般情况下由协议版本、状态码及其描述组成，其格式为 HTTP-Version Status-Code Reason-Phrase CRLF，其中 HTTP-Version 表示服务器 HTTP 协议的版本；Status-Code 表示服务器发回的响应状态代码；Reason-Phrase 表示状态代码的文本描述。状态代码由三位数字组成，第一个数字定义响应的类别，且有五种可能的取值，具体如下。

1xx：指示信息，表示请求已接收，继续处理。

2xx：成功，表示请求已被成功接收、理解、接受。

3xx：重定向，要完成请求必须进行更进一步的操作。

4xx：客户端错误，请求有语法错误或请求无法实现。

5xx：服务器端错误，服务器未能实现合法的请求。

常见响应状态代码如表 3-5 所示。

表 3-5 HTTP 返回状态值

响应状态代码	说明
200	服务器处理请求成功
301	请求的页面已经被永久移动到新的位置，服务器返回此响应，会自动跳转至新的位置
302	指出被请求的文档或页面已被临时移动到别处，用户应继续使用原有位置来进行之后的请求
401	未授权的请求，需要进行身份验证
403	服务器拒绝请求
404	服务器不存在该用户的请求资源
500	服务器出现错误异常，无法完成请求
501	服务器不具备完成请求的功能
505	服务器不支持请求中所用的 HTTP 协议版本

（2）响应头

响应头用来描述服务器的基本信息。服务器通过这些数据的描述可以通知客户端如何处理一会儿需要回送的数据。响应头包含服务器支持的请求、文字编码等内容。

（3）空行

空行用回车或者换行的标识进行内容的分割，用来告诉服务器响应头到此为止。

（4）响应体

响应体指服务器返回给客户端的文本信息。

2.response 对象实例

【实例】实现使用 response 对象动态显示当前时间。

使用 response 对象的 setHeader 方法动态显示当前时间，效果如图 3-14 所示。

现在的时间是:

2020-6-11 11:20:30

图 3-14　response 对象实例 1

通过设置 setHeader 方法，将间隔设置为 1 秒，实现动态改变事件，如示例代码 3-7 所示。

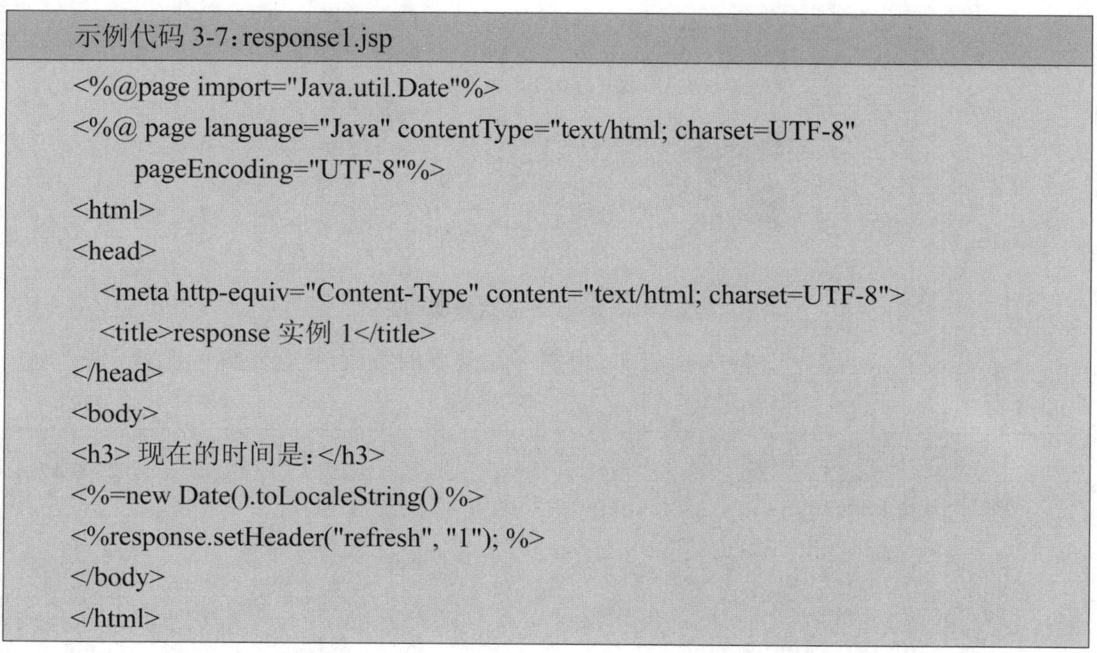

示例代码 3-7：response1.jsp

```
<%@page import="Java.util.Date"%>
<%@ page language="Java" contentType="text/html; charset=UTF-8"
    pageEncoding="UTF-8"%>
<html>
<head>
  <meta http-equiv="Content-Type" content="text/html; charset=UTF-8">
  <title>response 实例 1</title>
</head>
<body>
<h3> 现在的时间是：</h3>
<%=new Date().toLocaleString() %>
<%response.setHeader("refresh", "1"); %>
</body>
</html>
```

【实例】实现 response 重定向方法。

使用 response 对象中的 sendRedirect() 方法实现重定向。它与 request 对象的 forward 跳转的不同之处在于 sendRedirect() 方法会响应定向到新的 URL 且不能传递参数，跳转效果如图 3-15（a）所示，单击“确定”按钮之后，重定向至天津大学出版社官方网页，效果如图 3-15（b）所示。

图 3-15　response 对象实例 2

response2.jsp 页面为一个 form 表单，根据下拉菜单的数值，传递不同的参数，如示例代码 3-8 所示。

示例代码 3-8：response2.jsp

```
<%@ page language="Java" contentType="text/html; charset=UTF-8"
    pageEncoding="UTF-8"%>
<html>
<head>
  <meta http-equiv="Content-Type" content="text/html; charset=UTF-8">
  <title>response 对象实例 2</title>
```

```
</head>
<body>
网站友情链接：
<hr>
<form action="result.jsp" method="post">
  <select name="link">
    <option value="tjdxcbs" selected> 天津大学出版社 </option>
    <option value="jyb"> 中华人民共和国教育部 </option>
  </select>
  <input type="submit" name="submit" value=" 确定 ">
</form>
<hr>
</body>
</html>
```

通过 form 表单将下拉菜单的数值传送至 result.jsp 页面处理，判断传递过来的参数，根据不同的参数，使用 response 对象重定向至不同的页面，本实例中重定向至天津大学出版社官方网站，具体如示例代码 3-9 所示。

示例代码 3-9：result.jsp

```
<%@ page language="Java" contentType="text/html; charset=UTF-8"
    pageEncoding="UTF-8"%>
<html>
<head>
  <meta http-equiv="Content-Type" content="text/html; charset=UTF-8">
  <title>result 页面 </title>
</head>
<body>
<%
  String address = request.getParameter("link");
  if(address!=null)
  {
    if(address.equals("tjdxcbs"))
    {
      response.sendRedirect("http://www.tjupress.com.cn/");
    }
    else
      response.sendRedirect("http://www.moe.gov.cn/");
  }
```

```
    %>
    </body>
    </html>
```

技能点五 session 对象

1.session 对象概述

session 对象是与请求相关的 HttpSession 对象，它封装了特定用户会话所需的属性及配置信息。从一个客户打开浏览器并连接到服务器开始，到客户关闭浏览器离开服务器结束，为一个会话。当用户在应用程序的 Web 页面之间跳转时，存储在 session 对象中的变量不会丢失，而是在整个用户会话中一直存在下去。当用户请求来自应用程序的 Web 页面时，如果该用户还没有会话，则 Web 服务器将自动创建一个 session 对象。当会话过期或被放弃后，服务器将终止该会话。

在使用 session 对象的过程中，一般会用到获取指定属性变量值的方法，session 对象常用的方法见表 3-6。

表 3-6 session 对象常用方法

方法	说明
setAttribute(String name,Object obj)	在 session 中设定 name 所指定的属性值为 obj
getAttribute(String name)	返回 session 中 name 所指定的属性值
getAttributeNames()	返回 session 中所有变量的名称
removeAttribute(String name)	删除 session 中 name 所指定的属性
invalidate()	销毁与用户关联的 session
getCreationTime()	返回 session 建立的时间，返回值为 1970 年 1 月 1 日开始计算到 session 建立时的毫秒数
getLastAccessedTime()	返回客户端对服务器端提出请求至少处理 session 中数据的最后时间，若为新建的 session，则返回 -1
getMaxInactiveInterval()	获取 session 对象的生存时间
getId()	返回当前 session 对象的编号
isNew()	判断是否是一个新的 session

2. session 对象实例

【实例】实现 session 对象获取基本属性的常用方法。

运用 session 对象中的常用方法获取基本的属性，效果如图 3-16（a）所示。单击“请按这里”跳转至 test.jsp 页面处理 session 对象中的信息，效果如图 3-16（b）所示。

（a）

test页面
localhost:8080/JavaWeb/test.jsp
Test.jsp代码 获取session中的数据为：
您好，我们正在使用session对象传递数据！

（b）

图 3-16　session 对象实例

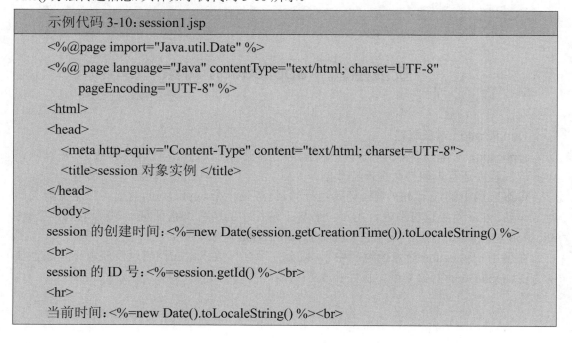

使用 session 对象常用方法 getCreationTime()、getId()、isNew() 获取信息，通过 setAttribute() 方法传递信息，具体如示例代码 3-10 所示。

示例代码 3-10：session1.jsp

```
<%@page import="Java.util.Date" %>
<%@ page language="Java" contentType="text/html; charset=UTF-8"
    pageEncoding="UTF-8" %>
<html>
<head>
  <meta http-equiv="Content-Type" content="text/html; charset=UTF-8">
  <title>session 对象实例 </title>
</head>
<body>
session 的创建时间：<%=new Date(session.getCreationTime()).toLocaleString() %>
<br>
session 的 ID 号：<%=session.getId() %><br>
<hr>
当前时间：<%=new Date().toLocaleString() %><br>
```

```
该 session 是新创建的吗？:<%=session.isNew() ? " 是 ": " 否 " %><br>
<hr>
<%session.setAttribute("info", " 您好，我们正在使用 session 对象传递数据！ "); %>
已向 Session 中保存数据，请单击下面的链接将页面重定向到 test.jsp
<a href="test.jsp"> 请按这里 </a>
</body>
</html>
```

test.jsp 页面使用 getAttribute() 方法获取 session 对象中的信息，具体如示例代码 3-11 所示。

示例代码 3-11：test.jsp

```
<%@ page language="Java" contentType="text/html; charset=UTF-8"
    pageEncoding="UTF-8"%>
<html>
<head>
  <meta http-equiv="Content-Type" content="text/html; charset=UTF-8">
  <title>test 页面 </title>
</head>
<body>
Test.jsp 代码 获取 session 中的数据为：<br>
<%=session.getAttribute("info") %>
</body>
</html>
```

技能点六 application 对象

1.application 对象概述

application 对象用于在多个程序中保存信息，可以在所有用户间共享信息，并可以在 Web 应用程序运行期间持久地保持数据。

Web 应用中的任何 JSP 页面开始执行时，将产生一个 application 对象。当服务器关闭时，application 对象也将消失。在同一个 Web 应用中的所有 JSP 页面，都将存取同一个 application 对象，即所有客户共享这个内置的 application 对象。

在使用 application 对象的过程中，一般会使用到存放指定属性值以及获取指定属性值的方法，application 对象常用方法如表 3-7 所示。

表 3-7　application 对象常用方法

方法	说明
int getMajorVersion()	取得主要的 Servlet API 版本
int getMinorVersion()	取得次要的 Servlet API 版本
String getServerInfo()	取得的名称和版本
String getMimeType(String file)	取得指定文件的 MIME 类型
String getRealPath(String path)	取得本地端 Path 的绝对路径
void log(String message)	将信息写入 log 文件中
void log(String message,Throwable throwable)	将 stack trace 所产生的异常信息写入 log 文件中
public void setAttribute(String name,Object object)	将数据保存到 application 对象
Object getAttribute(String name)	返回由 name 对象指定的 application 对象属性的值。该方法的使用与 session 对象相同
removeAttribute(String name)	从 application 对象中删除指定的属性

在 JSP 页面中，JSP 对象作用范围的对象分别为 page、request、session、application，它们之间的关系如图 3-17 所示。

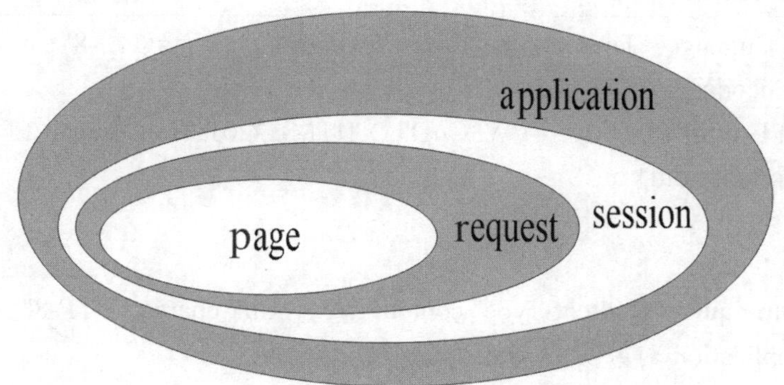

图 3-17　JSP 对象作用范围

page 只在当前页面有效，也就是在用户请求的页面有效，当前页面关闭或转到其他页面时，page 对象将在响应回馈给客户端后释放。

request 在当前请求中有效，可以通过 setAttribute() 方法实现页面中的信息传递，也可以通过 forward() 方法进行页面间的跳转。

session 在当前会话中有效，当同一浏览器对服务器进行多次访问时，在多次访问之间传递的信息就是 session 作用域的范围。

application 在所有的应用程序中都有效，也就是从服务器开始到服务器结束这段时间，application 作用域中存储的数据都是有效的，同样可以通过 setAttribute 赋值和 getAttribute 取值。

2.application 对象实例

【实例】实现使用 application 对象统计页面访问次数。

应用 application 的特性统计页面访问次数，页面每被访问一次数值加 1，效果如图 3-18 所示。

图 3-18 application 对象实例

使用 application 对象设置页面访问人数 count 属性过程如下。首次登录页面时，创建 count 属性，对该属性进行判断，若 application 中没有 count 属性，则为该属性赋值，若存在 count 属性，则将该属性值 +1 再进行赋值，达到统计网页访问人数的需求，具体如示例代码 3-12 所示。

示例代码 3-12：index.jsp

```
<%@ page language="Java" contentType="text/html; charset=UTF-8"
      pageEncoding="UTF-8" %>
<!DOCTYPE html PUBLIC "-//W3C//DTD HTML 4.01 Transitional//EN" "http://www.
w3.org/TR/html4/loose.dtd">
<html>
<head>
  <meta http-equiv="Content-Type" content="text/html; charset=UTF-8">
  <title>application 对象实例 </title>
</head>
<body>
<%
  // 获取 Object 对象
  String strNum = (String) application.getAttribute("count");
  int count = 0;
// 如果一个 Object 对象存在则说明有用户访问
  if (strNum != null) {
    count = Integer.parseInt(strNum) + 1;
  }
// 人数值加 1 后重新对 count 赋值
  application.setAttribute("count", String.valueOf(count));
%>
```

```
您是第 <%=application.getAttribute("count") %> 位访问者！
</body>
</html>
```

【实例】实现 application 对象常用方法。

应用 application 对象的部分常用方法，实现效果如图 3-19 所示。

```
←  →  C   ⓘ localhost:8080/JavaWeb/

JSP引擎名及Servlet版本号:Apache Tomcat/9.0.31
Java程序设计与项目实训教程
null
```

图 3-19　application 对象常用方法实例

应用 application 对象的基本方法 getServerInfo() 获取 Servlet 版本号、setAttribute() 设置属性值、getAttribute() 获取属性、removeAttribute() 移除属性方法，具体如示例代码 3-13 所示。

示例代码 3-13：applicationDemo.jsp

```
<%@ page language="Java" contentType="text/html; charset=UTF-8"
      pageEncoding="UTF-8" %>
<html>
<head>
  <meta http-equiv="Content-Type" content="text/html; charset=UTF-8">
  <title>application 对象实例 2</title>
</head>
<body>
JSP 引擎名及 Servlet 版本号：<%=application.getServerInfo() %>
<br>
<%
  application.setAttribute("name", "Java 程序设计与项目实训教程 ");
  out.print(application.getAttribute("name") + "<br>");
  application.removeAttribute("name");
  out.print(application.getAttribute("name") + "<br>");
%>
</body>
</html>
```

技能点七 其他内置对象

除了 out、request、response、session、application 对象之外，JSP 页面中还可以使用其余四个内置对象，即 config、page、pageContext 和 exception 对象。

1.config 对象

config 对象主要用于获得服务器的配置信息，初始化一些常用方法，如表 3-8 所示。

表 3-8 config 对象常用方法

方法	说明
ServletContext getServletContext()	返回所执行的 Servlet 的环境对象
String getServletName()	返回所执行的 Servlet 的名字
String getInitParameter(String name)	返回指定名字的初始参数值
Enumeration getInitParameterName()	返回该 JSP 中所有初始参数名

2.page 对象

page 对象是当前页面转换后的 Servlet 类的实例，可以从 JSP 脚本和表达式中获得一个内置对象。从转换后的 Servlet 类的代码中可以看到 Object page = this，只是 this 是引用的一个代名词。在 JSP 页面中，很少使用 page 对象。

3.pageContext 对象

pageContext 对象是一个比较特殊的对象，它的作用是取得任何范围的参数，通过 pageContext 对象可以获取 JSP 页面的 out、request、response、session、application 等对象，或者可以重新定向客户的请求等。

在使用 pageContext 对象的过程中，通常会使用到返回当前指定对象的方法，pageContext 对象的常用方法如表 3-9 所示。

表 3-9 pageContext 对象常用方法

方法	说明
forward(String url)	把页面转发到另一个页面或者 Servlet 组件
getAttribute(String.name[,int scope])	scope 参数是可选的，该方法用来检索一个特定的已经命名的对象属性值
getException()	返回当前的 exception 对象
getRequest()	返回当前的 request 对象
getResponse()	返回当前的 response 对象
getServletConfig()	返回当前页面的 Servletconfig 对象
setAttribute()	设置默认页面范围或特定对象范围之中的已命名对象
removeAttribute()	删除默认页面范围或特定对象范围之中的已命名对象

4.exception 对象

exception 对象用来处理 JSP 文件在执行时发生的错误和异常，可以配合 page 指令一起使用。在 page 指令中 isErrorPage 属性应设为 true，否则无法编译。

通过 exception 对象的方法指定某一个页面为错误处理页面，把所有的错误都集中在该页面进行处理，可以使整个系统的健壮性得到加强，也使程序的流程更加简单明晰。

技能点八　Cookie 对象

Cookie 是客户端技术，不属于内置对象，但是在 Web 项目中占有很大比重，与 session 对象有着密切的关系。

1.Cookie 对象概述

用户在浏览网站时，Web 服务器会将一些资料存放在客户端，这些资料包括用户在浏览网站期间输入的一些文字或选择记录。当用户下一次访问该网站的时候，服务器会从客户端查看是否有保留下来的 Cookie 信息，然后依据 Cookie 的内容，呈现特定的页面内容给用户。

Cookie 给网站和用户带来的好处非常多，具体包括以下几点。

1）Cookie 可以使得站点跟踪特定的访问者获取他们的访问次数、最后访问的时间以及进入站点的路径。

2）Cookie 可以获取站点上用户访问比较多的内容，方便统计，进而提供个性化服务。

3）Cookie 能够获取用户登录信息，在 Cookie 过期之前登录站点均无须用户名和密码，可以直接登录页面。

Cookie 带来了许多的便利，但是相对应地也会承担一些风险，具体如下。

1）潜在的安全性风险，Cookie 在存活时间内容易被恶意攻击，通过篡改 Cookie 致使用户在访问合法网站时被重定向至一些恶意网站。

2）Cookie 能够存储的信息非常有限，只有 4 kB 左右。

3）Cookie 可能被用户禁用，使得程序内使用到 Cookie 的部分失效。

Cookie 是存放在客户端浏览器的信息， session 是存放在服务器端的信息，两者之间的区别如表 3-10 所示。

表 3-10　Cookie 与 session 区别

比较内容	Cookie	session
存活时间	浏览器未关闭之前以及设定失效时间之内	浏览器未关闭之前（会话失效前）以及默认时间内
存在方式	客户机	服务器
数量限制	20（同一服务器）	无
处理速度	快	慢
保存内容	字符串	对象

2.Cookie 对象的常用方法

在使用 Cookie 对象的过程中，会使用到 Cookie 对象有效期、域名或者获取 Cookie 对象的有关方法，Cookie 对象常用方法如表 3-11 所示。

表 3-11 Cookie 对象常用方法

方法	说明
setDomain(String pattern)	设置 Cookie 的域名
getDomain()	获取 Cookie 的域名
setMaxAge(int expiry)	设置 Cookie 有效期，以秒为单位，默认当前 session 的存活时间
getMaxAge()	获取 Cookie 有效期，以秒为单位，默认为 -1，表明 Cookie 会存活到浏览器关闭为止
getName()	返回 Cookie 的名称，名称创建后将不能被修改
setValue(String newValue)	设置 Cookie 的值
getValue()	获取 Cookie 的值
setPath(String uri)	设置 Cookie 的路径，默认为当前页面目录下的所有 URL，还有此目录下的所有子目录
getPath()	获取 Cookie 的路径
setSecure(boolean flag)	指明 Cookie 是否要加密传输
setComment(String purpose)	设置注释描述 Cookie 的目的
getComment()	返回描述 Cookie 目的的注释，若没有则返回 null

3.Cookie 对象的基本操作

Cookie 是使用 key-value 形式来保存信息的。

（1）Cookie 的创建

调用 Cookide 对象的构造函数可以创建 Cookie。Cookie 对象的构造函数有两个字符串参数：Cookie 名和 Cookie 值。创建 Cookie 对象的语法如下：

```
Cookie cookieName=new Cookie(String key, String value)
```

（2）Cookie 的传送

在 JSP 页面中可使用 response 对象中的 addCookie() 方法将 Cookie 对象传送至客户端，具体的语法如下：

```
response.addCookie(cookieName)
```

（3）Cookie 的读取

JSP 通过 response 对象的 addCookie() 方法写入 Cookie 后，读取时将会调用 JSP 中 request 对象的 getCookies () 方法，该方法将会返回一个 Cookie 对象数组，因此必须通过遍历的方式进行访问。Cookie 通过 key-value 方式保存，因而在遍历数组时，需要通过调用 getName () 对每个数组成员的名称进行检查，直至找到需要的 Cookie，然后再调用 Cookie

对象的 getValue() 方法获得与名称对应的值。读取 Cookie 的语法如下：

```
Cookie[] cookies=request.getCookies()
```

例如：读取 Cookie 中的用户名，需要使用 for 循环进行遍历找出对应的用户名。

```
Cookie[] cookies=request.getCookies();
        String user="";
        for(int i=0;i<cookies.length;i++){
            if(cookies[i].getName().equals("user")){

    user=cookies[i].getValue();
            }
        }
```

4. Cookie 对象实例

【实例】实现使用 Cookie 对象保存并读取用户信息。

在首次登录页面时，输入用户名信息，实现注册，效果如图 3-20（a）所示。单击“确定”后，将填写的信息存入 Cookie，再次登录之后，会直接跳转至欢迎再次光临的页面，并显示注册的时间，效果如图 3-20（b）所示。

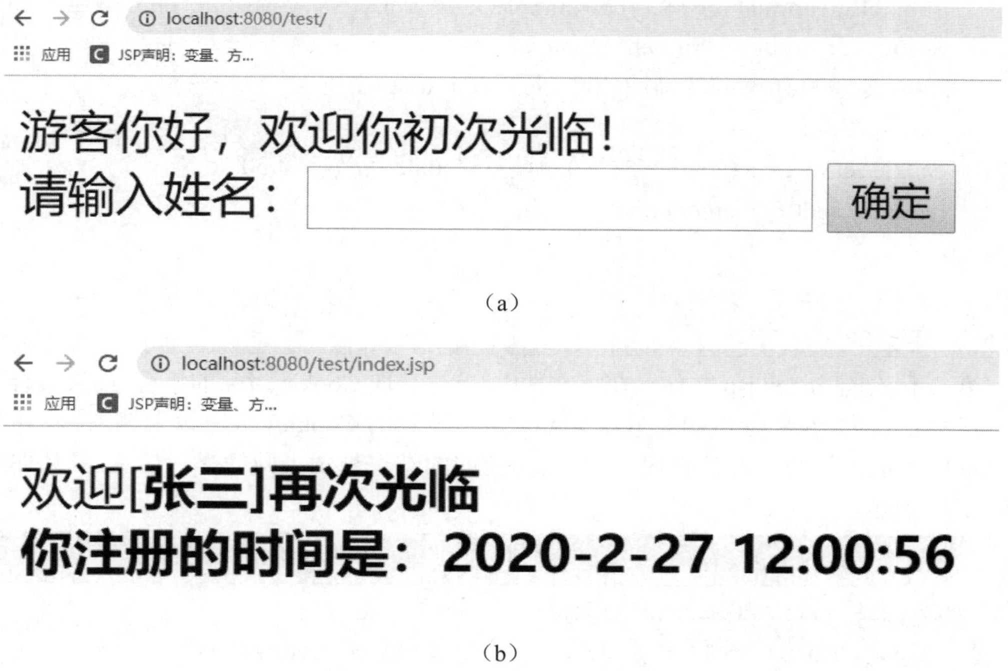

（a）

（b）

图 3-20　Cookie 对象实例

要实现图 3-20 效果，需要以下几个步骤。

第一步：编写 deal.jsp 文件，从前端获取用户名数据，用来向 Cookie 中写入注册信息，具体如示例代码 3-14 所示。

示例代码 3-14：deal.jsp

```
<%@page import="Java.text.SimpleDateFormat"%>
<%@page import="Java.util.Date"%>
<%@page import="Java.net.URLEncoder"%>
<%@ page language="Java" contentType="text/html; charset=UTF-8"
    pageEncoding="UTF-8"%>
<html>
<head>
  <meta http-equiv="Content-Type" content="text/html; charset=UTF-8">
  <title> 写入 Cookie</title>
</head>
<body>
<%
  request.setCharacterEncoding("UTF-8");// 设置请求的编译为 UTF-8
  String user=URLEncoder.encode(request.getParameter("user"),"UTF-8");// 获取用户名
  Date date = new Date();
  SimpleDateFormat format = new SimpleDateFormat("yyyy-MM-dd_hh:mm:ss");
  String currentTime = format.format(date);
  Cookie cookie =new Cookie("lee",user+"#"+currentTime);
// 创建并实例化 Cookie 对象 lee
  cookie.setMaxAge(60*60*24*30);// 设置 Cookie 有效期为 30 天
  response.addCookie(cookie);
%>
</body>
</html>
```

第二步：创建 index.jsp 文件。在该文件中，首先获取 Cookie 对象的集合，如果集合不为空，就通过 for 循环遍历 Cookie 集合，从中找出设置的 Cookie（这里设置为 lee），并从该 Cookie 中提取出用户名和注册时间，再根据获取的结果显示不同的提示信息，具体如示例代码 3-15 所示。

示例代码 3-15：index.jsp

```
<%@page import="Java.net.URLDecoder" %>
<%@ page language="Java" contentType="text/html; charset=UTF-8"
    pageEncoding="UTF-8" %>
<html>
<head>
  <meta http-equiv="Content-Type" content="text/html; charset=UTF-8">
```

```
    <title> 通过 cookie 保存并读取用户登录的信息 </title>
  </head>
  <body>
  <%
    Cookie[] cookies = request.getCookies();// 从 request 中获取 Cookie 对象的集合
    String user = "";// 登录用户
    String date = "";// 注册时间
    if (cookies != null) {
      for (int i = 0; i < cookies.length; i++)// 使用 for 循环遍历整体 Cookies
      {
        if (cookies[i].getName().equals("lee"))// 寻找 Cookie 名为 lee 的对象
        {
          user = URLDecoder.decode(cookies[i].getValue().split("#")[0]);// 获取用户名
          date = cookies[i].getValue().split("#")[1];// 获取注册时间，Cookie 对象中的值
均使用 # 进行连接
        }
      }
    }
    if ("".equals(user) && "".equals(date)) {
      // 如果没有注册
  %>
  游客你好，欢迎你初次光临！
  <form action="deal.jsp" method="post">
    请输入姓名：<input name="user" type="text" value="">
    <input type="submit" value=" 确定 ">
  </form>
  <%
  } else {
    // 已经注册
  %>
  欢迎 <b><%=user %>
  </b> 再次光临 <br>
  你注册的时间是：<%=date %>
  <%
    }
  %>
  </body>
  </html>
```

技能点九　EL 表达式和 JSTL 的使用方法

在页面中可以使用 JSP 脚本动态输出信息，但是单纯使用 JSP 脚本和 HTML 标签相结合，会显得代码结构混乱、可读性差，导致维护困难。基于这些原因，可以使用 EL 对 JSP 的输出进行优化。

1.EL 表达式

EL 是 JSP 2.0 增加的技术规范，是一种简单的语言，提供了在 JSP 中简化表达式的方法，目的是尽量减少 JSP 页面中的 Java 代码，使 JSP 页面的处理程序编写起来更加简捷，便于开发和维护。

（1）EL 的功能

1）可用于获取 JavaBean 属性。

2）能够读取集合类型对象中的元素。

3）可以使用运算符进行数据处理。

4）可以屏蔽一些常见的异常。

5）可以自动实现类型的转换。

（2）EL 表达式语法

```
${EL 表达式 }
```

在使用 EL 表达式获取变量前，必须将操作的对象保存在作用域中。${ userinfo} 代表获取变量 userinfo 的值。EL 提供“.”和“[ ]”两种操作符来存储数据。

1）一般采用 Java 代码一样的方式，使用“.”操作符来访问对象的属性。例如，${news.name} 可以访问 news 对象的 name 属性。

2）“[ ]”和“.”操作符类似，不仅可以用来访问对象的属性，还可以用来访问数字和集合。

【实例】使用 EL 表达式访问变量和集合，实现效果如图 3-21 所示。

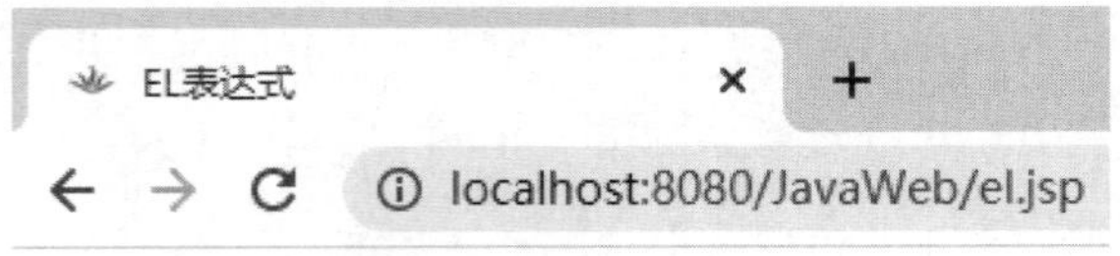

name信息：john

这是list集合中的第一条数据：EL

图 3-21　EL 表达式实例

使用 request.setAttribute() 方法将 name 属性和 list 集合置于作用域内，再使用 ${name} 和 ${list[0]} 方法在页面中进行调用，具体如示例代码 3-16 所示。

示例代码 3-16：el.jsp

```
<%@page import="Java.util.ArrayList" %>
<%@ page language="Java" contentType="text/html; charset=UTF-8"
      pageEncoding="UTF-8" %>
<%
   String name = "john";
   request.setAttribute("news", name);
   ArrayList list = new ArrayList();
   list.add("EL");
   list.add("Javaweb");
   request.setAttribute("list", list);
%>
<html>
<head>
   <meta http-equiv="Content-Type" content="text/html; charset=UTF-8">
   <title>EL 表达式 </title>
</head>
<body>
<p>name 信息：${news}</p>
<p> 这是 list 集合中的第一条数据：${list[0] }</p>
</body>
</html>
```

（3）EL 表达式的作用域访问对象

JSP 提供了四种作用域，分别是 page、request、session 和 application。EL 表达式也分别提供了四种作用域访问对象来实现数据的读取，如表 3-12 所示。

表 3-12　EL 表达式作用域访问对象

名称	说明
pageScope	用于获取页面范围（page）内的属性值
requestScope	用于获取请求范围（request）内的属性值
sessionScope	用于获取会话范围（session）内的属性值
applicationScope	用于获取应用程序范围（application）内的属性值

在使用 EL 表达式访问某个属性值时，应该指定查找范围，若没有指定查找范围则会按照 page、request、session、application 的顺序进行查找。例如使用 request 请求范围内的 news 对象中 name 属性的 EL 方法如下。

```
${ requestScope.news["name"] }
```

2.JSTL 的使用方法

EL 表达式在页面中的功能十分有限，不能进行逻辑处理。对于循环、判断等逻辑方法，还是需要与 Java 代码混合使用。而 JSTL 则可以实现逻辑控制，进一步优化代码。

JSTL（即 JSP 标准标签库）是由 JCP（Java Community Proces）所制定的标准规范，它主要提供给 Java Web 开发人员一个标准通用的标签库，并由 Apache 的 Jakarta 小组来维护。开发人员可以利用这些标签取代 JSP 页面上的 Java 代码，从而提高程序的可读性，降低程序的维护难度。

（1）JSTL 的准备工作

1）下载 JSTL 所需要的 jstl.jar 和 standard.jar 文件。

2）将两个 jar 文件粘贴到 WEB-INF\lib 目录下，并添加到项目中。

3）在 JSP 页面中添加标签指令（prefix 前缀可修改），具体如下：

<%@taglib uri="http://java.sun.com/jsp/jstl/core" prefix="c" %>

完成这三个步骤之后，就可以在添加标签指令的 JSP 页面使用 JSTL。

（2）JSTL 基本的使用方法（前缀设为 c）

JSTL 常用方法如表 3-13 所示。

表 3-13 JSTL 常用方法

方法	说明
<c:out>	用于在 JSP 中显示数据
<c:set>	用于保存数据
<c:remove>	用于删除数据
<c:catch>	用于处理产生错误的异常状况，并且将错误信息储存起来
<c:if>	用于逻辑判断
<c:choose>	本身只当作 <c:when> 和 <c:otherwise> 的父标签
<c:when>	<c:choose> 的子标签，用来判断条件是否成立
<c:otherwise>	<c:choose> 的子标签，接在 <c:when> 标签后，当 <c:when> 标签判断为 false 时被执行
<c:import>	检索一个绝对或相对 URL，然后将其内容暴露给页面
<c:forEach>	基础迭代标签，对集合的遍历
<c:forTokens>	根据指定的分隔符来分隔内容并迭代输出
<c:param>	用来给包含或重定向的页面传递参数
<c:redirect>	重定向至一个新的 URL
<c:url>	使用可选的查询参数来创造一个 URL

1）out 标签。value 属性是需要输出显示的表达式，对应的语法如下：

```
<c:out value=" "/>
```

2）forEach 标签。var 为集合中元素的名称，items 为集合对象，varStatus 为当前循环的状态信息，一般用于循环的索引号，对应的语法如下：

```
<c:forEach var="varName" items="items"  varStatus=" ">
    ….
</c:forEach>
```

3）forTokens 标签。var 为集合中元素的名称，items 为集合对象，delims 为分隔符，对应的语法如下：

```
<c:forTokens var="varName" items=" items " delims=" " >
    ….
</c:forTokens>
```

【实例】使用 JSTL 标签库中的 out 、forEach 和 forTokens 标签，实现循环输出数据，效果如图 3-22 所示。

图 3-22　JSTL 实例

使用 <c:out value=" "/> 显示对应的数据，forEach 标签、forTokens 标签进行循环输出，具体如示例代码 3-17 所示。

示例代码 3-17：jstl.jsp

```
<%@page import="Java.util.ArrayList" %>
<%@ page language="Java" contentType="text/html; charset=UTF-8"
    pageEncoding="UTF-8" %>
<!DOCTYPE html PUBLIC "-//W3C//DTD HTML 4.01 Transitional//EN" "http://www.w3.org/TR/html4/loose.dtd">
<%@taglib uri="http://Java.sun.com/jsp/jstl/core" prefix="c" %>
<%
```

```
        ArrayList list = new ArrayList();// 创建集合
        list.add("a");
        list.add("b");
        list.add("c");
        list.add("d");
        request.setAttribute("list", list);// 使用 add 方法添加到 list 集合中，并设置作用域
    %>
    <html>
    <head>
        <meta http-equiv="Content-Type" content="text/html; charset=UTF-8">
        <title>JSTL 实例 </title>
    </head>
    <body>
    使用 out 标签进行输出：<c:out value=" 这是 out 标签 "/>// 使用 out 标签进行输出
    <hr>
    forEach 标签 :<c:forEach var="news" items="${list }">// 将 list 集合设置为 items 属性，
并将该集合设置为 news 名称，在 forEach 标签中均使用 news 进行操作
        <c:out value="${news }"></c:out>
    </c:forEach>
    <br>
    <hr>
    forTokens 标签 :<br>
    <c:forTokens items=" 第一 , 第二 , 第三 " delims="," var="name">//items 设置为字符
串，delims 间隔符为 ",", 设置名称为 name
    <c:out value="${name}"/><p>
        </c:forTokens>
    </body>
    </html>
```

运用 JSP 内置对象、EL 表达式以及 JSTL 的有关知识完成网上书城项目主页设计，效果如图 3-23 所示。

（a）

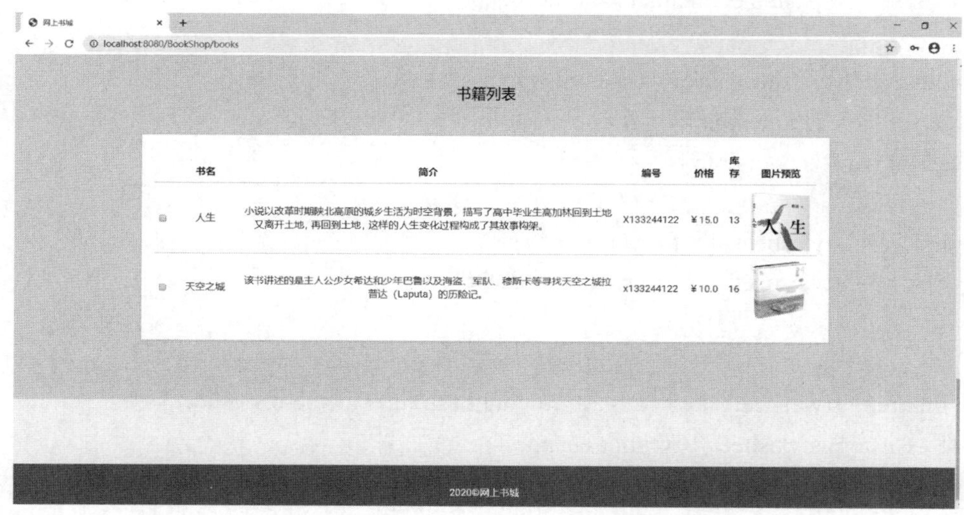

（b）

图 3-23　网上书城项目主页样式

第一步：在 web 文件夹下创建 main.jsp 页面，如图 3-24 所示。

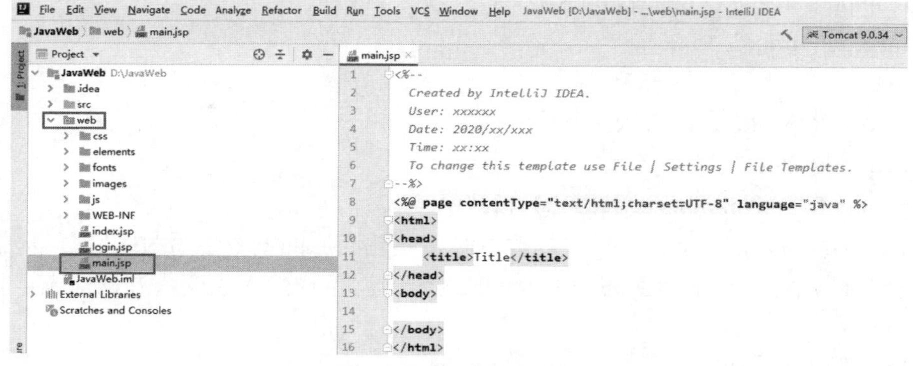

图 3-24　创建 main.jsp

第二步：在 main.jsp 页面中，应用轮播图样式，使用 include 动作标记引入头部文件、底部文件和菜单栏。使用 Map、List 集合法编写书籍的静态数据，并通过 request. setAttribute ("books",books) 方法获取编写的静态书籍信息，以便使用 EL 表达式和 JSTL 标签库的知识将图书信息，如书名、简介、编号、库存和价格等数据信息循环遍历在页面上。主页代码如示例代码 3-18 所示。

示例代码 3-18：main.jsp

```
<%@ page language="Java" import="java.util.*" contentType="text/html; charset=UTF-8" isELIgnored="false" %>
<%@page import="com.xt.util.PageTools" %>
<%@page import="com.xt.entity.Book" %>
<%@taglib uri="http://java.sun.com/jstl/core_rt" prefix="c" %>
<jsp:include page="elements/index_head.jsp"/>

<link rel="stylesheet" href="css/bootstrap.css">
<script src="js/jquery.js"></script>
<script src="js/bootstrap.js"></script>

<link rel="stylesheet" href="css/style4.css">
<link rel="stylesheet" href="css/icomoon.css">


<link rel="stylesheet" href="css/style.default.css" id="theme-stylesheet">
<!-- Custom stylesheet - for your changes-->
<link rel="stylesheet" href="css/custom.css">

<%
  Map map = new HashMap();// 编写书籍列表数据
  List<Map> books = new ArrayList<>();
  map.put("image", "images/book/book_31.png");
  map.put("price", 15.00);
  map.put("bid", 2);
  map.put("bookname", " 人生 ");
  map.put("stock", 13);
  map.put("booknumber", "X133244122");
  map.put("introduction", " 小说以改革时期陕北高原的城乡生活为时空背景，描写了高中毕业生高加林回到土地又离开土地，再回到土地，这样的人生变化过程构成了其故事构架。");
  books.add(map);
```

```
        Map map2 = new HashMap();
        map2.put("image", "images/book/tiankong.png");
        map2.put("price", 10.00);
        map2.put("bid", 1);
        map2.put("bookname", " 天空之城 ");
        map2.put("stock", 16);
        map2.put("booknumber", "x133244122");
        map2.put("introduction", " 该书讲述的是主人公少女希达和少年巴鲁以及海盗、军
队、穆斯卡等寻找天空之城拉普达（Laputa）的历险记。");
        books.add(map2);
        request.setAttribute("books", books);
    %>
    <body>
    <jsp:include page="elements/main_menu.jsp"/>
    <div id="carousel-example-generic" class="carousel slide center-block" data-ride="carou-
sel">
        <!-- Indicators -->
        <ol class="carousel-indicators">
            <li data-target="#carousel-example-generic" data-slide-to="0" class="active"></li>
            <li data-target="#carousel-example-generic" data-slide-to="1"></li>
            <li data-target="#carousel-example-generic" data-slide-to="2"></li>
            <li data-target="#carousel-example-generic" data-slide-to="3"></li>
            <li data-target="#carousel-example-generic" data-slide-to="4"></li>
        </ol>

        <!-- Wrapper for slides -->
        <div class="carousel-inner" role="listbox">
            <div class="item active">
                <img src="images/01.jpg" alt="...">
                <div class="carousel-caption">

                </div>
            </div>
            <div class="item">
                <img src="images/02.jpg" alt="...">
```

```
            <div class="carousel-caption">
                </div>
        </div>
            <div class="item">
                <img src="images/03.jpg" alt="...">
                <div class="carousel-caption">

                </div>
            </div>
            <div class="item">
                <img src="images/04.jpg" alt="...">
                <div class="carousel-caption">

                </div>
            </div>

        </div>

        <!-- Controls -->
          <a class="left carousel-control" href="#carousel-example-generic" role="button" da-
ta-slide="prev">
            <span class="glyphicon glyphicon-chevron-left" aria-hidden="true"></span>
            <span class="sr-only">Previous</span>
        </a>
         <a class="right carousel-control" href="#carousel-example-generic" role="button" da-
ta-slide="next">
            <span class="glyphicon glyphicon-chevron-right" aria-hidden="true"></span>
            <span class="sr-only">Next</span>
        </a>
      </div>
      <div id="fh5co-pricing" class="fh5co-bg-section">
        <div class="container">
            <div class="row animate-box">
                <div class="col-md-12 text-center fh5co-heading">
                <h2> 书籍列表 </h2>
```

```
        </div>
                </div>
        <div class="row">
            <div id="basket1" class="col-lg-12">
                <div class="box">
                    <div class="table-responsive">
                        <table class="table">
                            <thead>
                            <tr class="title">
                                <th class="checker"></th>
                                <th> 书名 </th>
                                <th> 简介 </th>
                                <th> 编号 </th>
                                <th class="price"> 价格 </th>
                                <th class="store"> 库存 </th>
                                <th class="view"> 图片预览 </th>
                            </tr>
                            </thead>
                            <tbody>
                             <c:forEach var="book" items="${books}">// 使用 JSTL 标签循环输出
书籍信息
                                <tr>
                                          <td><input type="checkbox" name="bookId" id="bookId" val-
ue="${book.bid}"/></td>
                                    <td class="title">${book.bookname}</td>
                                        <input type="hidden" name="title" value="${book.bid}:${book.
bookname}"/>
                                    <td>${book.introduction}</td>
                                    <input type="hidden" name="introduction"
                                         value="${book.bid}:${book.introduction}"/>
                                    <td>${book.booknumber}</td>
                                             <input type="hidden" name="booknumber" value="${book.
bid}:${book.booknumber}"/>
                                    <td> ￥${book.price}</td>
                                  price}"/>
```

```
                        <td>${book.stock}</td>
    <input type="hidden" name="stock" value="${book.bid}:${book.stock}"/>
                            <td class="thumb"><img src="${book.image}" width="100px"
height="100px"/></td>
                        <input type="hidden" name="image" value="${book.bid}:${book.
image}"/>
                    </tr>
                </c:forEach>
                </tbody>
            </table>

          </div>
        </div>
      </div>
    </div>
  </div>
</div>

</body>
<jsp:include page="elements/main_bottom.html"/>// 引入底部文件
<script type="text/javascript">
  function cart() {
    var input = document.getElementsByTagName('input');
    var countCheckBox = 0;
    var countChecked = 0;
    for (var i = 0; i < input.length; i++) {
      if (input[i].checked === true) {
        countChecked++;// 获取 checkbox 被勾上的数量
      }
    }
    if (countChecked == 0) {
      alert(" 请至少勾选一件商品 ");
      window.location.href = '/BookShop_Web_exploded/books';
    }
  }
```

第三步：引入头部文件，代码如示例代码 3-19 所示。

示例代码 3-19：index_head.jsp

```
<%@page language="Java" import="java.util.*" pageEncoding="UTF-8"%>
<!DOCTYPE html PUBLIC "-//W3C//DTD XHTML 1.0 Transitional//EN" "http://www.w3.org/TR/xhtml1/DTD/xhtml1-transitional.dtd">
<html xmlns="http://www.w3.org/1999/xhtml">
<head>
  <meta http-equiv="pragma" content="no-cache">
  <meta http-equiv="cache-control" content="no-cache">
  <meta http-equiv="expires" content="0">
  <meta http-equiv="Content-Type" content="text/html; charset=utf-8" />
  <title> 网上书城 </title>
  <link type="text/css" rel="stylesheet" href="css/style.css" />
</head>
```

第四步：引入菜单栏、订单模块、购物车模块以及搜索框功能，具体如示例代码 3-20 所示。

示例代码 3-20：main_menu.jsp

```
<%@ page language="Java" import="java.util.*" pageEncoding="UTF-8" isELIgnored="-false"%>
<%@ page import="com.xt.util.OnlineCounterListener" %>
<%@ page import="com.xt.util.OnlineCounter" %>
<!-- Bootstrap CSS -->
<link rel="stylesheet" href="css/bootstrap2.min.css">
<!-- FontAwesome CSS -->
<link rel="stylesheet" href="css/font-awesome.2min.css">
<!-- ElegantFonts CSS -->
<link rel="stylesheet" href="css/elegant-fonts.css">
<!-- themify-icons CSS -->
<link rel="stylesheet" href="css/themify-icons.css">
<!-- Swiper CSS -->
<link rel="stylesheet" href="css/swiper.min.css">
<!-- Styles -->
<!-- Styles -->
<link rel="stylesheet" href="css/style3.css">
<body></body>
<div class="hero-content">
    <header class="site-header">
```

```
<div class="top-header-bar">
<div class="container-fluid">
<div class="row">
<div class="col-12 col-lg-6 d-none d-md-flex flex-wrap justify-content-center justi-
fy-content-lg-start mb-3 mb-lg-0">
<div class="header-bar-email d-flex align-items-center">
<font color="BLACK"> 欢 迎 您，<strong>${loginuser}</strong></
font>   
</div>
<div class="header-bar-text lg-flex align-items-center">
<p> 在线人数：<%=OnlineCounter.getOnline()%></p>
        // 获取在线人数统计，运用 Servlet 监听器实现
</div>
</div>
<div class="col-12 col-lg-6 d-flex flex-wrap justify-content-center justify-con-
tent-lg-end align-items-center">
<div class="header-bar-search">
<form class="flex align-items-stretch" method="post" name="search" action="-
search">
<input type="search" placeholder=" 输入您想搜索的书籍 " name="keywords" >
<button type="submit" value="" class="flex justify-content-center align-items-center">
搜索 </button>
</form>
</div>
<div class="header-bar-menu">
<ul class="flex justify-content-center align-items-center py-2 pt-md-0">
<li><a onclick="window.location='register.jsp';" > 注册 </a></li>
<li><a onclick="window.location='logout.jsp';" > 登录 </a></li>
</ul>
</div>
</div>
</div>
</div>
</div>
<div class="nav-bar">
<div class="container">
<div class="row">
<div class="col-9 col-lg-3">
```

```
        <div class="site-branding">
        <h1  class="site-title"><a  href="/main.jsp"  rel="home">Line<span>Book</span></
a></h1>
        </div>
        </div>
        <div class="col-3 col-lg-9 flex justify-content-end align-content-center">
        <nav class="site-navigation flex justify-content-end align-items-center">
        <ul class="flex flex-column flex-lg-row justify-content-lg-end align-content-center">
        <li><a href="books"> 首页 </a></li>
        <li><a  href="showOrder?username=${loginuser}"> 我的订单 </a></li>// 跳转至订
单页面的按钮
        <li><a href="showCart"> 购物车 </a></li>// 跳转至购物车页面的按钮
        </ul>
        <div class="hamburger-menu d-lg-none">
        <span></span>
        <span></span>
        <span></span>
        <span></span>
        </div>
        <div class="header-bar-cart">
        <a href="#" class="flex justify-content-center align-items-center"></a>
        </div>
        </nav>
        </div>
        </div>
        </div>
        </div>
        </header>
    </div>
```

第五步：引入底部文件，代码如示例代码 3-21 所示。

示例代码 3-21：main_button.jsp

```
 <html>
<div class="footer">
   <div class="container">
      <div class="footer-main">
         <div class="container">
             <div class="copy-rights-main">
```

```
                <p>2020&copy; 网上书城 </p>
              </div>
            </div>
          </div>
        </div>
      </div>
    </html>
```

完成上述操作后，网上书城项目主页前端部分已经完成，效果如图 3-23 所示。

本次任务通过对 JSP 内置对象、Cookie 对象、EL 表达式以及 JSTL 的介绍，使读者重点熟悉 JSP 内置对象的常用方法、EL 表达式的应用方式、JSTL 的引入和使用，学习循环遍历数据的方法，了解 session 对象和 Cookie 对象之间的区别，为以后制作网上书城项目打下坚实的基础。

out	在外	request	请求
response	响应	session	会话
application	应用	config	配置
exception	异议		

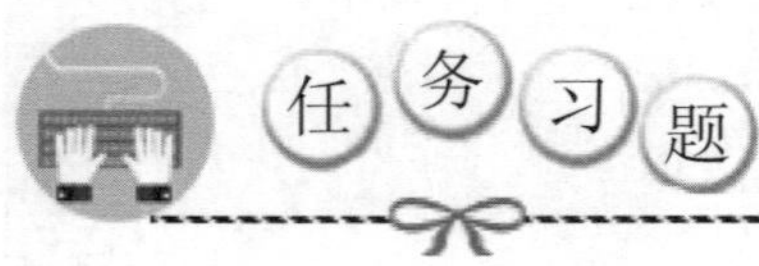

1. 选择题

1）要在 session 对象中保存属性，可以使用以下（　　）语句。

A. session.getAttribute（"key"，"value"）

B. session.setAttribute（"key"，"value"）

C. session.setAttribute（"key"）

D. session.getAttribute（"key"）

2）使用 request 对象的 getParameter 方法可以读取一个输入控件的（　　）属性值。

A .type　　B. value　　C. size　　D. name

3）能清除缓冲区中的数据，并且把数据写到客户端是 out 对象中的方法是 (　　)。

A. out.newLine()　B. out.clear()　C.out.flush()　D. out.clearBuffer()

4）以下陈述错误的是（　）。

A. 在 Web 项目的共享数据范围内，application 是范围最广泛的

B. 当我们在一个 JSP 页面新开窗口时，新开窗口的页面也共享 session 范围内的数据

C. 当在 JSP 页面中通过 <jsp:forward> 指令将页面请求转发到的页面中，可以共享一个 page 范围内的数据

D. 当用户重新打开一个浏览器窗口时，原 session 对象不再有效

5）给定 test1.jsp 代码片段，如下：

```
<html>
<jsp:include page="test2.jsp"flush="false">
<jsp:param name="color" value="red"/>
</jsp:include>
</html>
```

以下能获取参数“color”的是（　）。

A. <%=request.getParameter("color")%>

B. <%=request.getAttribute("color")%>

C. <jsp:getParamname="color"/>

D. <jsp:include param="color"/>

2. 填空题

1）JSP 主要内置对象有：_________、exception、pageContext、Cookie、request、response、________、out、config、page。

2）EL 表达式的语法：________。

3）在 JSP 中，request 内置对象代表请求消息，response 内置对象代表响应消息，_____ 内置对象代表会话。

4）从 HTTP 请求中，获得请求参数，应该调用 _________ 方法。

5）使用 response 对象进行重定向时，使用的方法是_________。

3. 简答题

简述 session 对象和 Cookie 对象的区别及它们各自的优势。

第四章　网上书城项目业务对象封装

通过学习 JavaBean 相关知识，了解 JavaBean 的概念，掌握 JavaBean 的编写规范以及种类，熟悉在 JSP 中使用 JavaBean 的方法，实现对于网上书城项目业务对象的封装。在任务实现过程中：

- 了解 JavaBean 的概念；
- 掌握 JavaBean 的编写规范；
- 掌握在 JSP 中使用 JavaBean 的方法；
- 掌握 JavaBean 的作用范围。

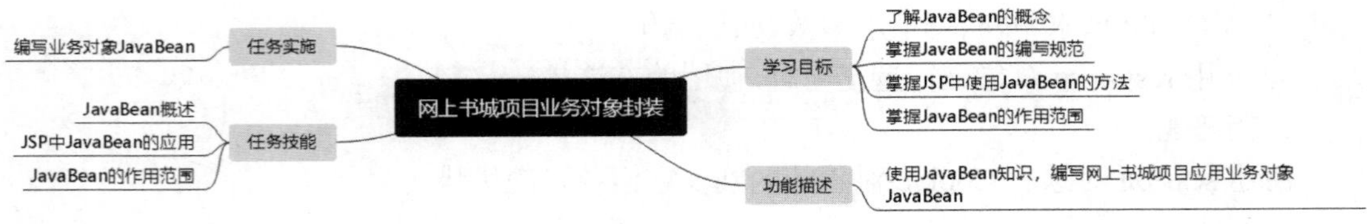

【情境导入】

在构建一个完整的项目时，对于实体属性的设置和获取是比较困难的。在页面和业务逻辑中如何快速、准确地获取实体属性是开发阶段的一大难题，JavaBean 的出现为开发人员提供了解决办法。

【功能描述】

● 使用 JavaBean 知识，编写网上书城项目应用业务对象 JavaBean。

技能点一　JavaBean 概述

1.JavaBean 由来

在 JavaWeb 开发的早期阶段，并没有逻辑分层的概念，只关注如何实现功能。如示例代码 4-1 所示，JSP 的 HTML 代码与 Java 业务逻辑代码混合在一起，这种结构混乱的代码书写方式，不能体现面向对象开发模式的优点，达不到代码的重用性，给程序的调试、维护以及后续开发带来了很大的困扰，直至 JavaBean 的出现，这一问题才得到了有效解决。

JavaBean 能够把复杂需求分析分解成简单的功能模块。这种模块是相对独立的部分，可以重用，为 JavaWeb 开发提供了一个极好的解决方案。

示例代码 4-1：早期开发代码示例

```
<body>
// 业务逻辑代码
 <%!class Person{
  String name;
  int age;
  Person(String name, int age) {XM=xm; AGE=a;}
  String getName() { return this.name;}
}%>
 <%Person p=new Person(" 小李 ",18);%>
//HTML 网页代码
   姓名：<%=p.getName()%>
</body>
```

2.JavaBean 简介

JavaBean 是一种可重用组件，它是由 Java 语言编写的，可以被 Servlet、Applet、JSP 等应用程序调用，也可以被 Java 开发工具使用。通过提供设计模式的方法，在系统中其他 Java 类通过反射机制、get 和 set 方法获取对应的属性值，并对这些属性值进行赋值和修改操作。这种与 HTML 代码分离，使用 Java 代码编写封装的类，就是一个 JavaBean。

JavaBean 在发展的过程中不断改进，使其具有如下优势：

1）用户可以使用 JavaBean 将创造的对象进行打包；

2）可以简单重用，并应用到系统中的 Java 类；

3）提高了代码的可重用性；

4）增强了系统的可维护性。

3.JavaBean 编写规范

JavaBean 是一种遵循特定写法的 Java 类，是可重用的组件，它的方法命名、构造及行为必须符合特定的规范，具体内容如下：

1）提供一个默认的无参构造函数；

2）类变量为 private（私有）类型；

3）每一个类变量提供一组 public 类型的 getter、setter 方法。

在 JavaBean 对象中，为了防止外部直接对 JavaBean 属性的调用，通常将 JavaBean 中的属性设置为 private 类型，并通过两个 public 方法设置和调用该属性的值。

当对类变量编写赋值方法时，方法的命名规则为 set 属性名。其中，属性名的第一个字母应大写，方法详细代码如下：

```
public void setName(dataType data)
```

当对名为 Name 的类变量编写取值方法时，方法的命名规则为 get 属性名。其中，属性名的第一个字母应大写，方法详细代码如下：

```
public datatype getName()
```

4.JavaBean 种类

JavaBean 根据其实现内容的不同分为以下两种类型。

1）可视化 JavaBean：主要用于传统的 Java Swing 应用中，即图形界面程序。此类 JavaBean 封装了图形用户界面（GUI）的组件，如文本框、按钮、分隔窗格和表等。

2）非可视化 JavaBean：主要用于 JavaWeb 应用中。此类 JavaBean 用于封装应用的业务对象和业务逻辑代码。

下面通过实例来了解一下非可视化的 JavaBean。

【实例】将示例代码 4-1 中的业务逻辑代码改写为非可视化 JavaBean，类中封装入 name 和 age 属性，并提供获取和设置属性的方法 getter 和 setter，具体如示例代码 4-2 所示。

示例代码 4-2：非可视化 JavaBean

```
// 业务对象 JavaBean
public class Person {
   // 业务对象属性
   private String name;
private int age;
// 获取和设置业务对象属性的方法
   public String getName() { return name; }
   public void setName(String name) { this.name = name; }
```

```
    public int getAge() { return age; }
    public void setAge(int age) { this.age = age; }
}
```

技能点二　JSP 中 JavaBean 的应用

JavaBean 技术可以和 JSP 技术相互结合，实现商业逻辑层和表现层的分离，提高 JSP 程序的运行效率和重用程度。

1.JSP 中 JavaBean 的工作过程

JavaBean 将 JSP 页面中的 HTML 代码与 Java 代码分别编写，把业务逻辑代码单独封装成一个 Java 类，然后在 JSP 页面中调用此类。通过 JavaBean 可以降低 HTML 代码与 Java 代码之间的耦合度，简化 JSP 页面，使 Java 代码表现出面向对象语言的灵活性以及重用性，也使得后续的开发维护工作更加便利。在 Java Web 开发中，通常用 JavaBean 组件来完成业务对象的封装和业务逻辑的处理，如图 4-1 所示。

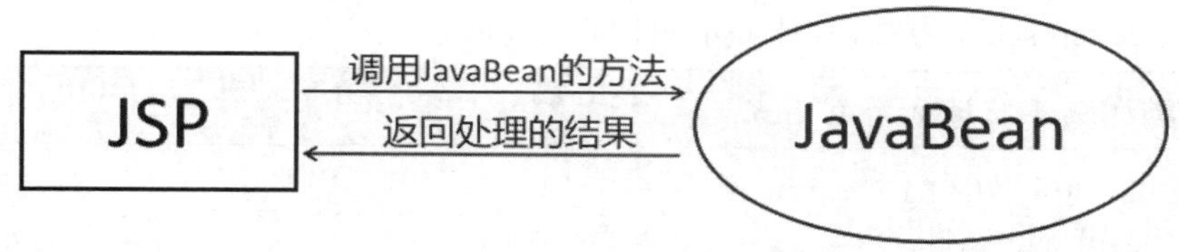

图 4-1　JavaBean 工作过程

2.JSP 中使用 JavaBean

JSP 页面为了集成和支持 JavaBean，提供了三个动作元素来访问 JavaBean，分别为 <jsp:useBean>、<jsp:getProperty> 和 <jsp:setProperty>。

（1）<jsp:useBean> 标签

<jsp:useBean> 标签主要用于在 JSP 中声明并创建一个 JavaBean 实例，声明后 JavaBean 对象实例转变为脚本变量，可以通过脚本元素或其他自定义标签来访问。其语法格式如下：

```
<jsp:useBean id="JavaBean 的名字 " class="JavaBean 的完整类名 " scope="bean 的作用域 " />
```

<jsp:useBean> 标签中的属性功能如表 4-1 所示。

表 4-1　<jsp:useBean> 标签属性功能

属性名	属性功能
id	JavaBean 的对象标识，每一个 id 对应一个唯一的 JavaBean 实例，可自主命名，在 JSP 页面中通过该属性来获取 JavaBean 实例

续表

属性名	属性功能
class	JavaBean 对象的完整类名，由包名 + 类名组成，编写时要注意大小写一致
scope	JavaBean 对象的作用域，其值可以是 page、request、session 或 application，该属性默认值为 page

【实例】编写 JavaBean 对象，并在网页中输出其属性的值，输出效果如图 4-2 所示。

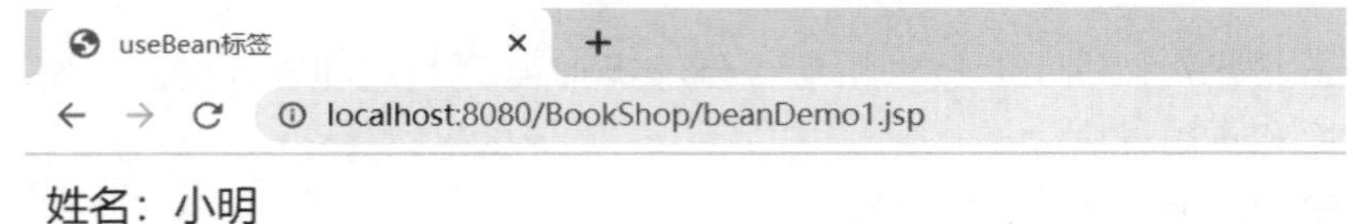

姓名：小明

年龄：18

图 4-2 <jsp:useBean> 标签实例

为实现如图 4-2 所示效果，在 JavaBean 中定义两个属性 name 和 age，并分别为其赋值，然后配置 getter、setter 方法，具体如示例代码 4-3 所示。

示例代码 4-3：JavaBean 对象

```
public class Person{
  //JavaBean 属性
private String name = " 小明 ";
private int age = 18;
// 属性 getter、setter 方法
  public String getName() { return name; }
  public void setName(String name) { this.name = name; }
  public int getAge() { return age; }
  public void setAge(int age) { this.age = age; }
}
```

在 JSP 页面中使用 <jsp:useBean> 标签实例化 Bean 对象，并使用 EL 表达式输出对象的 name 和 age 属性值，具体如示例代码 4-4 所示。

示例代码 4-4：beanDemo1.jsp

```
<%@ page contentType="text/html;charset=UTF-8" language="java" %>
<html>
<head>
  <title>useBean 标签 </title>
```

```
</head>
<body>
// 实例化 JavaBean
<jsp:useBean id="person" class="javaBean.Person "></jsp:useBean>
//EL 表达式根据 JavaBean 的 id 获取其中的属性
<p> 姓名：${person.name}</p>
<p> 年龄：${person.age}</p>
</body>
</html>
```

（2）<jsp:getProperty > 标签

<jsp:getProperty> 标签主要用于在 JSP 页面获取 JavaBean 实例的属性值，并将其转换为 java.lang.String 对象，放置在隐含的 out 对象中，输出到 JSP 页面。使用该标签前须先创建 JavaBean 实例，语法格式如下：

```
<jsp:getProperty name="JavaBean 的名字 " property="JavaBean 属性名 "/>
```

<jsp: getProperty > 标签中的属性功能如表 4-2 所示。

表 4-2　<jsp: getProperty > 标签属性功能

属性名	属性功能
name	取值目标 JavaBean 实例的 id 值
Property	取值目标属性的属性名称

【实例】编写 JavaBean 对象，并在网页中输出其属性的值，输出效果如图 4-3 所示。

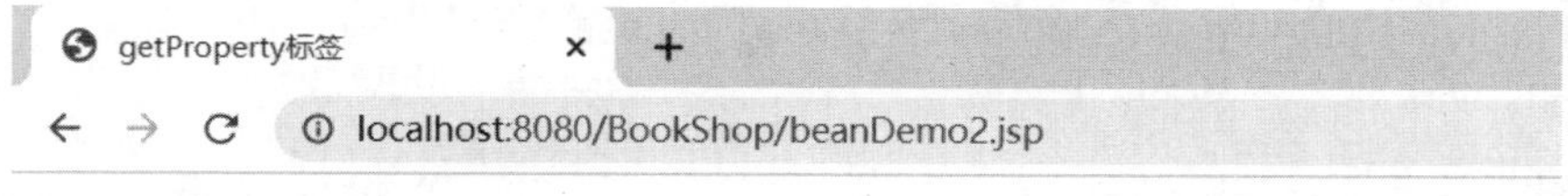

图 4-3　<jsp: getProperty > 标签实例

为实现如图 4-3 所示效果，在 JavaBean 中定义三个属性 name、price 和 stock，并分别为其赋值，然后配置 getter 和 setter 方法，具体如示例代码 4-5 所示。

示例代码 4-5：JavaBean 对象

```
public class Book {
```

```
private String name = " 算法导论 ";
private Double price = 56.00d;
private int stock = 10;
public String getName() { return name; }
public void setName(String name) { this.name = name; }
public Double getPrice() { return price; }
public void setPrice(Double price) { this.price = price; }
public int getStock() { return stock; }
public void setStock(int stock) { this.stock = stock; }
}
```

在 JSP 页面中使用 <jsp:useBean> 标签实例化 Bean 对象，并使用 <jsp: getProperty > 标签获取 Bean 中属性值并输出，具体如示例代码 4-6 所示。

示例代码 4-6：beanDemo2.jsp

```
<%@ page contentType="text/html;charset=UTF-8" language="java" %>
<html>
<head>
<title>getProperty 标签 </title>
</head>
<body>
<jsp:useBean id="book" class=" javaBean.Book">
</jsp:useBean>
<p> 书名：<jsp:getProperty property="name" name="book"/></p>
<p> 价格：<jsp:getProperty property="price" name="book"/> ￥</p>
<p> 库存：<jsp:getProperty property="stock" name="book"/></p>
</body>
</html>
```

（3）<jsp:setProperty > 标签

<jsp:setProperty > 标签主要用于设置 JavaBean 实例的属性值，使用该标签前须先创建 JavaBean 实例，语法格式如下：

```
<jsp:setProperty name="JavaBean 的 id"  property="JavaBean 属性名 " value=" 属性值 "/>
```

<jsp: setProperty > 标签中的属性功能如表 4-3 所示。

表 4-3 <jsp: setProperty > 标签属性功能

属性名	属性功能
name	赋值目标 JavaBean 实例的 id 值

续表

属性名	属性功能
property	赋值目标属性的属性名称
value	想要为目标属性赋的值，如该参数为空，则不会修改目标属性的属性值
param	页面请求 request 对象中的参数名，将该名称的参数的值赋给 JavaBean 的目标属性，相当于 value="<%=request.getParameter("paramValue")%>"，该属性不能和 value 同时使用

【实例】编写 JavaBean 对象，并在网页中输出其属性的值，输出效果如图 4-4 所示。

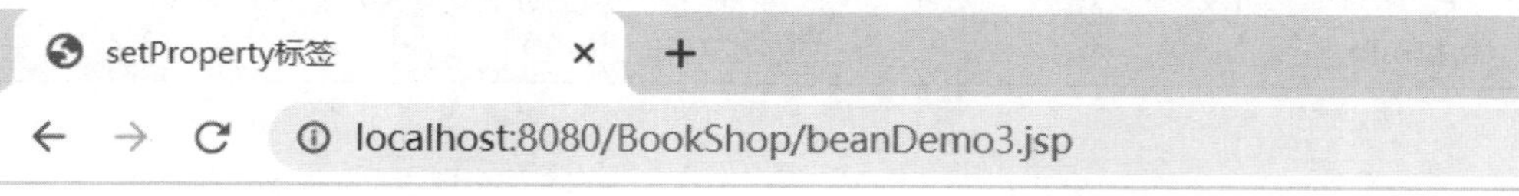

课程名称：C语言

分数：92

图 4-4　<jsp: setProperty > 标签实例

为实现如图 4-4 所示效果，在 JavaBean 中定义两个无初始值的属性 courseName 和 score，然后配置 getter 和 setter 方法，具体如示例代码 4-7 所示。

示例代码 4-7：JavaBean 对象

```
public class Course {
private String courseName;
private int score;
public String getCourseName() { return courseName; }
public void setCourseName(String courseName) { this.courseName = courseName; }
public int getScore() { return score; }
public void setScore(int score) { this.score = score; }
}
```

在 JSP 页面中使用 <jsp:useBean> 标签实例化 Bean 对象，使用 <jsp: setProperty> 标签为实例对象的属性赋值，并使用 <jsp: getProperty> 标签在页面中输出赋值后的属性，具体如示例代码 4-8 所示。

示例代码 4-8：beanDemo3.jsp

```
<%@ page contentType="text/html;charset=UTF-8" language="java" %>
```

```
<html>
<head>
  <title>setProperty 标签 </title>
</head>
<body>
<jsp:useBean id="course" class="com.xt.entity.Course"/>
<jsp:setProperty property="courseName" name="course" value="C 语言 "/>
<jsp:setProperty property="score" name="course" value="92"/>
<p> 课程名称:<jsp:getProperty name="course" property="courseName"/></p>
<p> 分数:<jsp:getProperty name="course" property="score"/></p> </body>
</html>
```

技能点三 JavaBean 的作用范围

通过设定 <jsp:useBean> 的 scope 属性，可使 JavaBean 组件对于不同的任务有不同的生命周期和不同的使用范围。scope 属性具有四种可供选择的属性值，分别是 Page、Request、Session 和 Application，其中 Page 为其默认值。JavaBean 组件只有在它定义的范围内才能使用，在它的活动范围外将无法访问。

（1）Page

Page 的有效范围是用户请求访问的当前 JSP 页面，在该范围内：

1）JavaBean 的实例对象存储在 PageContext 对象中；

2）每次请求访问 JSP 页面，都会创建一个新的 JavaBean 对象。

（2）Request

Request 的有效范围是任何具有相同请求的 JSP 文件，直到页面执行完毕向客户端发回响应或在此之前已通过某种方式（如重定向、链接等方式）转到另一个文件为止，在该范围内：

1）JavaBean 实例对象作为属性对象保存在 HttpRequest 对象中；

2）属性名为 <jsp:useBean> 标签设置的 id 值；

3）属性值是实例化后的 JavaBean 对象；

4）通过 HttpRequest.getAttribute() 方法通过 id 值获取 JavaBean 对象。

（3）Session

Session 的有效范围是 Session 的整个生存周期，在该范围内：

1）JavaBean 实例对象作为属性保存在 HttpSession 对象中；

2）属性名为 <jsp:useBean> 标签设置的 id 值；

3）属性值是实例化后的 JavaBean 对象；

4）对该 JavaBean 属性的任何改动，都会影响在此 Session 内其他 Page 对象和 Request

对象对该 JavaBean 的调用；

5）可通过 HttpSession.getAttribute() 方法通过 id 属性获取 JavaBean 对象。

（4）Application

Application 的有效范围是 Application 的整个生存周期，在整个 JavaWeb 应用的生命周期内都是有效的，在该范围内：

1）JavaBean 实例对象作为属性保存在 Application 对象中；

2）对该 JavaBean 属性的任何改动，都会影响在此 Application 内其他 JSP 内置对象对该 JavaBean 的调用；

3）可通过 application.getAttribute() 方法通过 id 属性获取 JavaBean 对象。

【实例】通过调用 JavaBean 实例在 JSP 页面中显示当前时间，并修改 JSP 页面中 JavaBean 实例的 scope 属性显示不同的结果，效果如图 4-5 所示。

图 4-5　JavaBean 显示时间实例

具体步骤如下。

第一步：创建名为 Time 的 Java 类，在类中添加时间 dateTime 属性和星期 week 属性。针对这两个属性编写获取日期和星期的 getter 方法。其中，getDateTime 方法通过 SimpleDateFormat 对象格式化当前时间，并将格式化后的时间字符串赋值给 dateTime 属性；getWeek 方法通过获取 Calendar 对象中当前星期所代表的 int 值，选择 weeks 数组中的当前星期字符串赋值给 week 属性。由于该类主要用于获取当前时间，并不涉及对两个属性的手动赋值，所以实例中不需要提供 setter 方法。具体如示例代码 4-9 所示。

示例代码 4-9：Time.java

```
public class Time {
    // 时间
    private String dateTime;
    // 星期
    private String week;
    private Calendar calendar = Calendar.getInstance();
    public String getDateTime() {
        Date date = calendar.getTime();
        SimpleDateFormat sdf = new SimpleDateFormat("yyyy 年 MM 月 dd 日 HH 点 mm 
分 ss 秒 ");// 对当前时间格式化为“年月日时分秒”形式
        dateTime=sdf.format(date);
```

```
        return dateTime;// 返回当前时间
    }
    public String getWeek() {
        String[] weeks = {" 星期日 "," 星期一 "," 星期二 "," 星期三 "," 星期四 "," 星期五 "," 星期六 "};
        week = weeks[calendar.get(Calendar.DAY_OF_WEEK)-1];// 获取当前星期
        return week;// 返回当前星期
    }
}
```

第二步：创建名为 time.jsp 的页面，在该页面中实例化 Time 对象，获取并输出 dateTime 属性和 week 属性，具体如示例代码 4-10 所示。

示例代码 4-10：time.jsp

```
<%@ page contentType="text/html;charset=UTF-8" language="java"%>
<html>
<head>
  <title>Time</title>
</head>
<body>
<jsp:useBean id="time" class="javaBean.Time" scope="page" />
<p><jsp:getProperty name="time" property="dateTime"/></p>// 当前时间
<p><jsp:getProperty name="time" property="week"/></p>// 当前星期
</body>
</html>
```

第三步：修改 Time 对象中的 scope 值，观察页面的显示结果。

1）当 scope=application 时，第一次访问 time.jsp，页面显示出系统当前时间。当进行刷新页面、重启浏览器、使用其他设备浏览等操作时，它显示的时间始终不变，皆为第一次访问 JSP 页面的时间（即 Bean 刚创建时得到的系统时间），因为 scope=application，所以 JavaBean 的实例在服务器内存中只有一份，此时只要不重新启动 Web 服务，输出不会变化，如图 4-6 所示。

图 4-6 scope 属性值为 application

2）当 scopp=session 时，浏览 time.jsp，进行刷新操作时显示也不会变化，如图 4-7 所示。而当重新打开浏览器访问 time.jsp 页面，即创建一个新的 session 会话时，服务器会再次创建 JavaBean 的实例，取得当前系统时间，这时将会得到访问的时间，如图 4-8 所示。

图 4-7　scope 属性值为 session 时首次打开浏览器

图 4-8　scope 属性值为 session 时重新打开浏览器

3）当 scope=page/request 时，刷新页面将不断销毁和创建新的 JavaBean 对象，得到的都是即时的系统时间，如图 4-9 和图 4-10 所示。

图 4-9　scope 属性值为 request 时刷新前

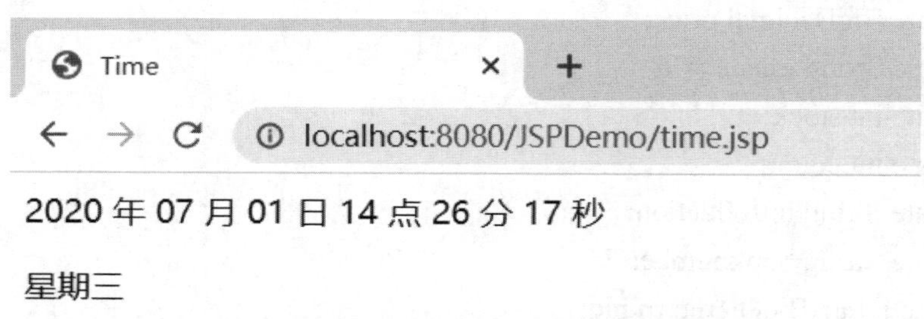

图 4-10　scope 属性值为 request 时刷新后

运用学习到的 JavaBean 知识，根据第一章网上书城需求分析和数据库设计，封装项目的四个业务对象：Book、Item、Order 和 UserInfo，具体步骤如下。

第一步：在 src 文件夹下创建 com.xt 包，在其下创建四个 Java 类，分别为 Book、Item、Order 和 UserInfo，效果如图 4-11 所示。

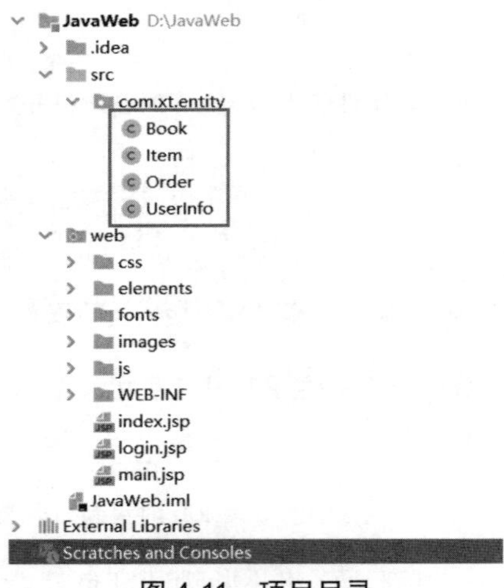

图 4-11 项目目录

第二步：编写封装书籍信息属性的名为 Book 的 JavaBean 对象，该对象有 8 个属性，分别是 int 类型的 bid、stock 和 count 属性，String 类型的 bookname、image、introduction 和 booknumber 属性，BigDecimal 类型的 price 属性，具体如示例代码 4-11 所示。

示例代码 4-11：Book.java

```
package com.xt.entity;
import java.math.BigDecimal;
public class Book {
    private int bid;
    private String bookname;
    private BigDecimal price;
    private String image ;
    private int stock;
    private int count;
    private String introduction;
    private String booknumber;
    public int getBid() {return bid;}
```

```
    public String getBookname() {return bookname;}
    public BigDecimal getPrice() {return price;}
    public String getImage() {return image;}
    public int getStock() {return stock;}
    public int getCount() {return count;}
    public String getIntroduction() {return introduction;}
    public String getBooknumber() {return booknumber;}
    public void setBid(int bid) {this.bid = bid;}
    public void setBookname(String bookname) {this.bookname = bookname;}
    public void setPrice(BigDecimal price) {this.price = price;}
    public void setImage(String image) {this.image = image;}
    public void setStock(int stock) {this.stock = stock;}
    public void setCount(int count) {this.count = count;}
    public void setIntroduction(String introduction) {this.introduction = introduction;}
    public void setBooknumber(String booknumber) {this.booknumber = booknumber;}
}
```

第三步：编写封装订单相关属性的名为 Item 的 JavaBean 对象，该对象有 7 个属性，分别是 int 类型的 iid、oid、bid、count 和 state 属性，String 类型的 createDate 属性，BigDecimal 类型的 dealPrice 属性，具体如示例代码 4-12 所示。

示例代码 4-12：Item.java

```
package com.xt.entity;
import java.math.BigDecimal;
public class Item {
    private int iid;
    private int oid;
    private int bid;
    private int count;
    private BigDecimal dealPrice;
    private String createDate;
    private int state;
    public int getIid() {return iid;}
    public void setIid(int iid) {this.iid = iid;}
    public int getOid() {return oid;}
    public void setOid(int oid) {this.oid = oid;}
    public int getBid() {return bid;}
    public void setBid(int bid) {this.bid = bid;}
    public int getCount() {return count;}
```

```
    public void setCount(int count) {this.count = count;}
    public BigDecimal getPrice() {return dealPrice;}
    public void setPrice(BigDecimal price) {this.dealPrice = price;}
    public String getCreateDate() {return createDate;}
    public void setCreateDate(String createDate) {this.createDate = createDate;}
            public int getState() {return state;}
    public void setState(int state) {this.state = state;}
}
```

第四步：编写封装订单模块相关属性的名为 Order 的 JavaBean 对象，该对象有 4 个属性，分别是 int 属性的 oid 和 uid，String 类型的 createDate 和 BigDecimal 类型的 total_price，具体如示例代码 4-13 所示。

示例代码 4-13：Order.java

```
package com.xt.entity;
import java.math.BigDecimal;
import java.util.Date;
public class Order {
    private int oid ;
    private int uid ;
    private BigDecimal total_price;
    private String createDate;
    public int getOid() {return oid;}
    public void setOid(int oid) {this.oid = oid;}
    public int getUid() {return uid;}
    public void setUid(int uid) {this.uid = uid;}
    public BigDecimal getTotal_price() {return total_price;}
    public void setTotal_price(BigDecimal total_price) {this.total_price = total_price;}
    public String getCreateDate() {return createDate;}
    public void setCreateDate(String createDate) {this.createDate = createDate;}
}
```

第五步：编写封装登录注册属性的名为 UserInfo 的 JavaBean 对象，该对象有 4 个属性，分别是 int 类型的 uid，String 类型的 username、password 和 email，具体如示例代码 4-14 所示。

示例代码 4-14：UserInfo.java

```
package com.xt.entity;
public class UserInfo {
    private int uid;
    private String username;
```

```
    private String password;
    private String email;
    public int getUID() { return uid; }
    public void setUID(int uid) { this.uid = uid; }
    public String getUsername() {return username;}
    public void setUsername(String username) {this.username = username;}
    public String getPassword() {return password;}
    public void setPassword(String password) {this.password = password;}
    public String getEmail() {return email;}
    public void setEmail(String email) {this.email = email;}
}
```

本次任务通过对 JavaBean 知识的学习，重点熟悉 JavaBean 封装方法，学习在 JSP 页面中使用 JavaBean，了解 JavaBean 的作用范围，封装网上书城项目的业务对象，为以后制作网上书城项目打下良好的基础。

entity	实体	introduction	介绍
attribute	属性	format	格式化
simple	简单	calendar	日历

1. 选择题

1）关于 JavaBean 正确的说法是（　　）。

A.Java 文件与 Bean 所定义的类名可以不同，但一定要注意区分字母的大小写

B. 在 JSP 文件中引用 Bean，其实就是用 <jsp:useBean> 语句

C. 被引用的 Bean 文件的文件名后缀为 .java

D.Bean 文件放在任何目录下都可以被引用

2）在 JSP 中调用 JavaBean 时不会用到的标记是（　　）。

A.<javabean>　　B.<jsp:useBean>

C.<jsp:setProperty> D.<jsp:getProperty>

3）在 JSP 中使用 <jsp:getProperty> 标记时，不会出现的属性是（　　）。

A.name B.property C.value D. 以上皆不会出现

4）在项目中已经建立了一个 JavaBean，该类为 bean.Student, 给 bean 赋予 name 属性，则下面标签用法正确的是（　　）。

A.<jsp:useBeanid="student" class="Student" scope="session"></jsp:useBean>

B.<jsp:useBeanid="student" class="Student" scope="session">hellostudent!</jsp:useBean>

C.<jsp:useBeanid="student" class="bean. Student" scope= "session">hello student!</jsp:useBean>

D.<jsp:getProperty name= ="name" property="student"/>

5）如果使用标记 <jsp:getProperty name="bean Name" property="property Name" /> 准备取出 bean 的属性值，但 property Name 属性在 beanName 中不存在，也就是说在 beanName 中没有这样的属性名，也没有 getPropertyName() 方法，那么会在浏览器中显示（　　）。

A. 错误页面 B.null C.0 D. 什么也没有

2. 操作题

编写一个 JavaBean，定义一个包，将该 Bean 编译后生成的类放入该包中，设计三个属性 name、age 和 sex。

第五章　网上书城项目数据库连接

通过学习 JDBC 的相关知识，掌握 JDBC 的基本概念和连接对象，掌握 JDBC 操作数据库的方法，能够运用 JDBC 编写网上书城项目连接数据库的操作，并实现增加、删除、修改和查询数据的功能。在任务实现过程中：

- 了解 JDBC 的概念和访问方式；
- 掌握 JDBC 操作数据库的流程；
- 掌握 JDBC 执行增加、删除、修改和查询操作；
- 掌握 JDBC 元数据、批处理和事务的使用方法。

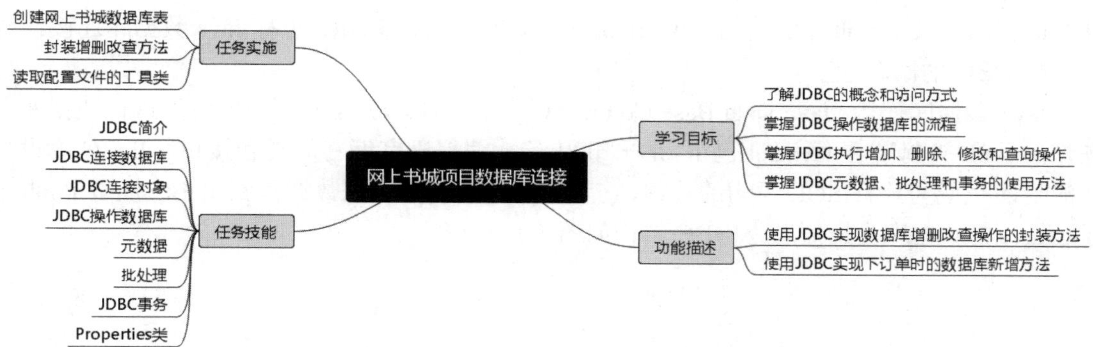

【情境导入】

开发实际软件项目时，总是需要存储数据，例如编写网站的注册登录功能时，注册的用户信息需要被保存在服务器的数据库中，在用户执行登录操作时，查询数据库中的用户信息，并跟登录操作输入的数据进行验证，从而实现登录的安全性验证。

【功能描述】

- 使用 JDBC 实现数据库增删改查操作的封装方法。
- 使用 JDBC 实现下订单时的数据库新增方法。

技能点一　JDBC 简介

软件开发中存在着著名的 2—8 原则，即 80 % 的软件都需要访问数据库。所以，大多数开发语言都提供了数据库的访问，Java 中访问数据库时使用 JDBC，也称 Java 数据库连接。

1.JDBC 的概念

Java 数据库连接（Java Data Base Connectivity），简称 JDBC，它是一种用于执行数据库操作的 API，被封装在 java.sql 包中，是一组用 Java 编程语言编写的类和接口。JDBC 库设计的目的是成为执行 SQL 语句的接口，它允许大型应用程序把数据写到 JDBC 接口上，再通过接口与数据库交换信息。JDBC 结构如图 5-1 所示。

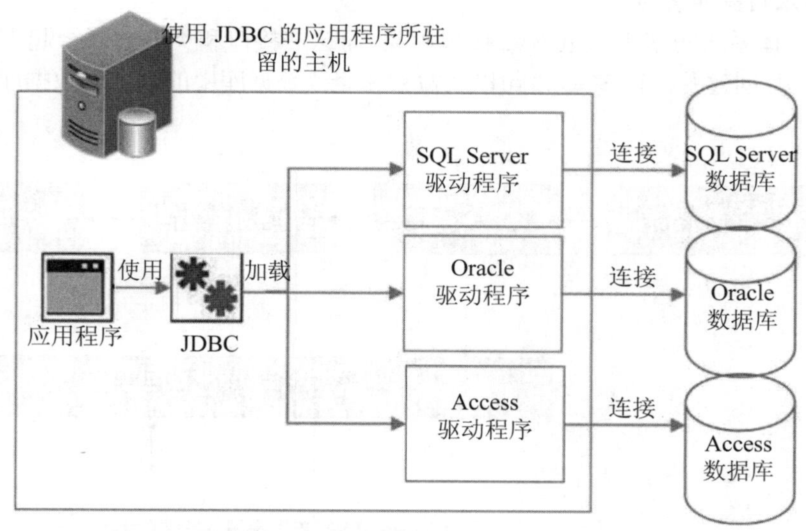

图 5-1　JDBC 结构

2.JDBC 的四类驱动

一个完整的 JDBC 应用由驱动程序、驱动程序管理器、应用程序三部分组成，具体内容如下。

（1）驱动程序

JDBC 的驱动程序实际是由 Sun 公司或数据库厂商提供的包，负责完成应用程序和特定的数据库通信。例如 MySQL 提供的 JDBC 驱动程序，就完成对 MySQL 的通信。JDBC 提供了四种类型的驱动程序。

1）JDBC-ODBC 桥。这类驱动程序将所有数据库调用通过 JDBC-ODBC 桥传送给 ODBC 连接，再由 ODBC 调用本地数据库驱动代码。该驱动连接方式一般只用于原型开发，而不用于正式的运行环境。

2）本地 API。本地 API 驱动直接把 JDBC 调用转变为数据库的标准调用，然后再去访问数据库。这类驱动程序没有使用纯 Java 的 API，需要把 Java 应用连接到数据源时，往往必须执行一些额外的配置工作。

3）JDBC 网络纯 Java 驱动程序。网络驱动协议是基于 Server 的，其访问数据库的方法为网络协议驱动——中间件服务器——数据库 Server。

4）本级协议纯 Java 驱动程序。本地驱动完全由 Java 实现，直接把 JDBC 调用转换为符合相关数据库系统规范的请求。

（2）驱动程序管理器

驱动程序需要在驱动程序管理器中注册，由驱动程序管理器组织应用程序连接指定类型的数据库。

（3）应用程序

应用程序一般用来表示完成某项或多项特定工作的计算机程序，主要运行在用户模式，能够很好地和用户进行交互，具有良好的用户界面。

3. 基本数据库访问

JDBC 体系结构基于一组 Java 接口和类，而这些接口和类能够联合起来使编程人员连接数据库源，创建并执行 SQL 语句以及检索和修改数据库中的数据。数据库操作流程如图 5-2 所示。

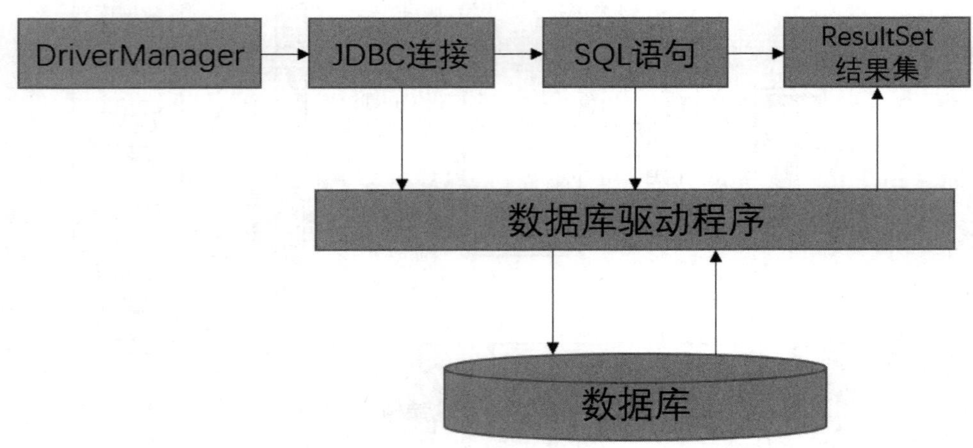

图 5-2　数据库操作流程

图 5-2 中的每个方框代表一个 JDBC 类或接口，这些类或接口在存取关系数据库的过程中提供了必要的方法。数据存取的一切工作从 DriverManager（驱动程序管理器）类开始，该类负责借助 JDBC 驱动程序建立与数据库的连接。

JDBC 数据驱动程序由实现 Driver（驱动程序）接口的类来定义。JDBC 驱动程序针对某个特定数据库的 SQL 请求进行转换，如果没有合适的驱动程序，就无法连接该数据库，因此加载连接 Java 类与特定数据库进行通信的驱动程序是实现一个应用所必须做的事情。一个基础的 JDBC 操作流程包含下列步骤。

第一步：引入必要的类。

第二步：加载 JDBC 驱动程序。

第三步：标识数据源。

第四步：使用驱动程序管理器创建 JDBC 连接对象 Connection。

第五步：使用连接对象创建 Statement 对象。

第六步：使用 Statement 对象执行 SQL 语句。

第七步：从执行返回的 ResultSet 对象中检索数据。

第八步：关闭 ResultSet 对象。

第九步：关闭 Statement 对象。

第十步：关闭 Connection 对象。

技能点二　JDBC 连接数据库

1. 数据库驱动程序注册

连接数据库前，需要将数据库厂商提供的数据库驱动类注册到 JDBC 的驱动程序管理器中，用来告知系统将访问哪种数据库，通常情况下是通过将数据库驱动类加载到 JVM 来实现的。加载时可以使用 Class 类的 forName 静态方法，把驱动程序注册到 DriverManager 中；或者是利用 JDBC DriverManager 类的 registerDriver() 方法。

当 MySQL 驱动程序使用 MySQL-connector-java 5 及以下版本进行 JDBC 连接时，注册语句如下：

```
Class.forName("com.MySQL.jdbc.Driver");
```

使用 MySQL-connector-java 6 及以上版本进行 JDBC 连接时，注册语句如下：

```
Class.forName("com.MySQL.cj.jdbc.Driver");
```

2. 连接数据库

当数据库驱动注册到 DriverManager 后，就可以利用 DriverManager 的 getConnection() 静态方法建立起一条客户机到数据库服务器的数据连接，在该方法中必须提供数据库连接字符串和相应的登录数据库的口令和密码，代码如下：

```
Connection con = DriverManager.getConnection(url,login_name,login_password);
```

其中，URL 是由数据库厂商制定的，不同的数据库，它的 URL 也有所区别，但都符合一个基本的格式，即“JDBC 协议 +IP 地址”或“域名 + 端口 + 数据库名称”。MySQL 数据库 URL 的语法格式如下：

```
protocol//[hosts][/database][?properties]
```

其中：

1）protocol 为 JDBC 协议，MySQL 的基础连接协议为 jdbc:MySQL；

2）hosts 为 IP 地址或域名 + 端口；

3）database 为要连接的数据库的名称；

4）properties 为 MySQL 连接的全局属性，格式为 key=value 的键值对形式，并用“&”符号分隔不同的属性。

在编写 MySQL 数据库的 URL 时，一般会使用 useUnicode 属性和 characterEncoding 属性指定数据库存储数据时的编码。在使用 MySQL-connector-java 6 及以上版本进行 JDBC 连接时，需要在全局属性中加入 serverTimezone 时区属性，例如在网上书城中，连接 URL 代码如下：

```
jdbc:MySQL://localhost:3306/bookshop?useUnicode=true&characterEncoding=utf8&-
serverTimezone=GMT%2B8
```

【实例】注册 MySQL-JDBC 驱动，连接 MySQL 数据库并在控制台中输出连接对象，关键代码如示例代码 5-1 所示。

示例代码 5-1：连接数据库

```
package jdbc;
import java.sql.Connection;
import java.sql.DriverManager;
import java.sql.SQLException;
public class JDBCDemo1 {
    public static void main(String[] args){
        try {
            Class.forName("com.MySQL.cj.jdbc.Driver");
            String url = "jdbc:MySQL://localhost:3306/ bookshop? useUnicode=true&"
                    + " characterEncoding=utf8&serverTimezone=GMT%2B8";
            String user = "root";
            String password = "123456";
            Connection conn = DriverManager.getConnection(url, user, password);
            System.out.println(" 连接成功，连接对象为 "+conn);
        } catch (ClassNotFoundException e) {
            System.out.println(" 驱动类获取失败 ");
            e.printStackTrace();
        }catch (SQLException e1){
            System.out.println("JDBC 连接失败 ");
            e1.printStackTrace();
        }
    }
}
```

运行结果如图 5-3 所示。

```
"C:\Program Files\Java\jdk1.8.0_251\bin\java.exe" ...
连接成功，连接对象为com.mysql.cj.jdbc.ConnectionImpl@481a15ff

Process finished with exit code 0
```

图 5-3 数据库连接成功效果图

技能点三　JDBC 连接对象

1.Connection 对象接口

Connection 接口位于 java.sql 包中，是数据库连接对象的接口，在项目中只有成功实例化该对象，才能访问数据库，并对数据库执行相关的增删改查操作，Connection 接口的方法如表 5-1 所示。

表 5-1　Connection 接口方法

方法名	方法说明
void close() throws SQLException	断开连接，释放 Connection 对象的数据库和 JDBC 资源
Statement createStatement() throws SQLException	创建一个 Statement 对象，将 SQL 语句发送到数据库
void commit() throws SQLException	用于提交 SQL 语句，确认从上一次提交 / 回滚以来进行的所有更改
boolean isClosed() throws SQLException	用于判断 Connection 对象是否已经被关闭
CallableStatement prepareCall(String sql) throws SQLException	创建一个 CallableStatement 对象，调用数据库存储过程
PreparedStatement prepareStatement (String sql) throws SQLException	创建一个 PreparedStatement 对象，将参数化的 SQL 语句发送到数据库
void rollback() throws SQLException	用于取消 SQL 语句，取消在当前事务中进行的所有更改

2.Statement 对象接口

在创建了数据库连接之后，可通过调用 SQL 语句对数据库进行操作，JDBC 中的 Statement 接口封装了这些操作。Statement 接口提供了执行 SQL 语句和获取操作结果的基本方法，如表 5-2 所示。

表 5-2　Statement 接口方法

方法名	方法说明
boolean execute(String sql) throws SQLException	执行指定的 SQL 语句。如果 SQL 语句返回结果，该方法返回 true，否则返回 false
ResultSet executeQuery(String sql) throws SQLException	执行查询类型（select）的 SQL 语句，该方法返回查询所获取的结果集 ResultSet 对象
executeUpdate int executeUpdate(String sql) throws SQLException	执行 SQL 语句中 DML 类型（insert、update、delete）的 SQL 语句，返回更新所影响的行数
void close() throws SQLException	立即释放 Statement 对象的数据库和 JDBC 资源

续表

方法名	方法说明
boolean isClosed() throws SQLException	判断 Statement 对象是否已被关闭，如果被关闭，则不能再调用该 Statement 对象执行 SQL 语句，该方法返回布尔值
void addBatch(String sql) throws SQLException	将 SQL 语句添加到 Statement 对象的当前命令列表中，该方法用于 SQL 命令的批处理
int[] executeBatch() throws SQLException	将一批 SQL 命令提交给数据库执行，返回更新计数组成的数组
void clearBatch() throws SQLException	清空 Statement 对象中的命令列表
ResultSet getGeneratedKeys() throws SQLException	检索由执行此 Statement 对象而创建的任何自动生成的主键

使用 Connection 对象的 createStatement 方法可以创建一个 Statement 对象。Statement 对象中存储了要执行的 SQL 语句，根据 SQL 语句种类的不同，使用 Statement 类中的不同方法运行，具体如下：

1）如果执行的是 select 语句，使用 executeQuery 方法；

2）如果执行的是 insert、update 或者 delete 语句，则使用 executeUpdate 方法；

3）如果预先不知道要执行的 SQL 语句类型，使用 execute 方法， execute 方法可以用于执行 select、insert、update 或者 delete 语句。

3.PreparedStatement 对象接口

Statement 接口封装了 JDBC 执行 SQL 语句和获取执行结果的方法，它可以完成 Java 程序执行 SQL 语句的操作，但在实际开发过程中， SQL 语句往往需要将程序中的变量作为 SQL 语句中的条件参数。使用 Statement 接口进行参数拼接操作过于烦琐，而且存在 SQL 注入等安全方面的缺陷。针对这一问题，JDBC 提供了 Statement 接口的子类 PreparedStatement，以便于执行参数化的 SQL 语句。

PreparedStatement 拥有 Statement 接口中的所有方法，而且 PreparedStatement 接口针对参数化 SQL 语句的执行进行了方法扩展。保存在 PreparedStatement 接口中的 SQL 语句，可以使用占位符“?”来代替 SQL 语句中的参数，然后调用方法对占位符进行赋值。在实际的开发过程中，基本都使用 PreparedStatement 接口进行 SQL 语句的执行，这样既可以提高代码编写效率，又能保证一定的安全性。PreparedStatement 接口的方法如表 5-3 所示。

表 5-3 PreparedStatement 接口方法

方法名	方法说明
void setString(int parameterIndex, String x) throws SQLException	将 String 值 x 作为 SQL 语句中的参数值，parameterIndex 为参数位置的索引
void setInt(int parameterIndex, int x) throws SQLException	将 int 值 x 作为 SQL 语句中的参数值，parameterIndex 为参数 位置的索引

续表

方法名	方法说明
void setLong(int parameterIndex, long x) throws SQLException	将 long 值 x 作为 SQL 语句中的参数值，parameterIndex 为参 数位置的索引
void setDouble(int parameterIndex, double x) throws SQLException	将 double 值 x 作为 SQL 语句中的参数值，parameterIndex 为 参数位置的索引
void setByte(int parameterIndex, byte x) throws SQLException	将 byte 值 x 作为 SQL 语句中的参数值，parameterIndex 为参 数位置的索引
void setFloat(int parameterIndex,float x) throws SQLException	将 float 值 x 作为 SQL 语句中的参数值，parameterIndex 为参 数位置的索引
void setObject(int parameterIndex, Object x) throws SQLException	将 Object 对象 x 作为 SQL 语句中的参数值，parameterIndex 为参数位置的索引
void setShort(int parameterIndex, short x) throws SQLException	将 short 值 x 作为 SQL 语句中的参数值，parameterIndex 为参 数位置的索引
void setBoolean(int parameterIndex,boolean x) throws SQLException	将布尔值 x 作为 SQL 语句中的参数值，parameterIndex 为参 数位置的索引
void setDate(int parameterIndex, Date x) throws SQLException	将 java.sql.Date 值 x 作为 SQL 语句中的参数值，parameterIndex 为参数位置的索引
void setTimestamp(int parameterIndex, Timestamp x) throws SQLException	将 Timestamp 值 x 作为 SQL 语句中的参数值，parameterIndex 为参数位置的索引

对于数据库中的主键一般采用自动递增的方式，为了使 prepareStatement 对象执行带有主键的 SQL 语句后接收自动生成的主键，需要在 Connection 对象创建 prepareStatement 实例时，传入相关参数以启动该功能，该步骤相关代码如下：

```
PreparedStatement ps =
 connection. prepareStatement(sql,Statement.RETURN_GENERATED_KEYS);
```

使用该 prepareStatement 对象实例执行带有主键的 SQL 语句后，通过调用 prepareStatement 对象的 getGeneratedKeys() 方法获得含有自增主键的 ResultSet 对象，遍历该 ResultSet 实例对象，获取第一行第一列的值即是返回的自增主键。如果此 prepareStatement 对象未生成任何键，则返回空的 ResultSet 对象，该过程关键代码如下：

```
prepareStatement.executeUpdate();
ResultSet rs = ps.getGeneratedKeys();
if(rs.next()){  int id = rs.getInt(1);  }
```

4.ResultSet 对象接口

在使用 Statement 接口或 PreparedStatement 执行查询语句后，会返回数据库中的查询结果，对此，JDBC 提供了 ResultSet 对象来封装返回的结果集。ResultSet 接口中提供了相关方法来操纵获取结果集中的记录。ResultSet 是基于光标功能来定位查询结果的，通过光标可以自由获取某一行中的数据，其方法如表 5-4 所示。

表 5-4 ResultSet 接口方法

方法名	方法说明
boolean next() throws SQLException	将光标位置向后移动一行，如移动的新行有效返回 true，否则返回 false
boolean previous() throws SQLException	将光标位置向前移动一行，如移动的新行有效返回 true，否则返回 false
int getInt(String columnLabel) throws SQLException	以 int 的方式获取 ResultSet 对象当前行中指定列的值，参数 columnLabel 为列名称
String getString(String columnLabel) throws SQLException	以 String 的方式获取 ResultSet 对象当前行中指定列的值，参数 columnLabel 为列名称
void close() throws SQLException	立即释放 ResultSet 对象的数据库和 JDBC 资源
boolean isClosed() throws SQLException	判断当前 ResultSet 对象是否已关闭
boolean first() throws SQLException	将光标移动到 ResultSet 对象的第一行
boolean last() throws SQLException	将光标移动到 ResultSet 对象的最后一行
boolean absolute(int row) throws SQLException	将光标移动到 ResultSet 对象的给定行编号，参数 row 为行编号

ResultSet 对象主要提供以下三种方法来获取数据。

（1）当前结果集指针移动方法

调用 next() 方法移动指针到想要访问的记录上，它的使用受记录及类型的限制。将当前记录指针移动到下一条记录上。每次获得的记录集，在访问具体记录前都必须执行这一方法，next 使当前记录指针定位到记录集的第一条记录。

（2）当前记录字段值获取方法

getTypeName() 方法用于读取当前记录指定字段的值， TypeName 为该字段类型名称，在此字段用 Java 的数据类型描述，而非数据类型描述，如取字符串类型字段是 getString() 而非 getVarchar()。Java 程序使用一组与 MySQL 类型不同的类型来代表值，但是两者之间互相兼容，这就使 Java 和 MySQL 能够交换以它们各自类型存储的数据，兼容的类型映射如表 5-5 所示。

表 5-5 Java 和 MySQL 可兼容的类型映射

MySQL 类型	Java 类型
char	String/Byte
varchar	String
int	Integer
bigint	Long
decimal	BigDecimal
double	Double
float	Float

续表

MySQL 类型	Java 类型
text	String
Datetime	Date

访问当前记录中的字段时，可以用字段名表明你所要访问的字段，也可以用该字段在发出查询的 select 子句中的字段位置表明所要访问的字段，两种方式的代码如下。

```
String name=rs.getString("ename"); // 通过字段名访问
String name=re.getString(2); // 通过字段位置访问
```

（3）更新当前字段值的方法

updateTypeName 方法，用于更新当前记录的指定字段的值。但是该方法的使用受语句对象的类型制约，该方法有两种更新方式，代码如下。

```
Rs.update("ename","zengcobra"); // 通过字段名更新，zengcobra 为更新后的字段名
Rs.update(2,"zengcobra"); // 通过字段位置更新，zengcobra 为更新后的字段名
```

技能点四　JDBC 操作数据库

1. 添加数据

通过 JDBC 向数据库添加数据，可以使用 INSERT 语句实现插入数据的 SQL 语句，对于 SQL 语句中的参数可以用占位符“？”代替，然后通过 PreparedStatement 对其赋值并执行 SQL，执行完毕后关闭连接资源，编写添加方法步骤详解如下。

第一步：注册 MySQL-JDBC 驱动，使用 DriverManager 类创建 JDBC 连接对象，核心代码如下。

```
Class.forName("com.mysql.cj.jdbc.Driver");
String url =
"jdbc:mysql://localhost:3306/
bookshop?useUnicode=true&characterEncoding=utf8&serverTimezone=GMT%2B8";
String user = "root";
String password = "123456";
Connection conn = DriverManager.getConnection(url, user, password);
```

第二步：编写添加数据的 INSERT 语句，自增的 uid 字段传入 NULL，其他字段使用占位符“？”代替，关键代码如下。

```
String sql = "INSERT INTO userinfo VALUES (NULL,?,?,?)";
```

第三步：使用 PreparedStatement 对象执行 SQL 语句，并通过 PreparedStatement 对象为 SQL 语句中的参数逐一赋值，赋值时要注意参数的下标值不是从 0 开始，而是从 1 开始计

数，与数组的下标有所区别。赋值完成后，调用 PreparedStatement 对象的 executeUpdate() 方法执行添加操作，此时新增的用户数据才写入到数据库中，该方法执行后返回 int 型数据，其含义是本次操作所影响的数据行数。该过程的关键代码如下。

```
// 初始化 prepareStatement
PreparedStatement ps = conn.prepareStatement(sql);
// 通过 PreparedStatement 对象为 SQL 语句中的参数逐一赋值
ps.setString(1,"user1");
ps.setString(2,"123456");
ps.setString(3,"123@edu.com");
// 调用 PreparedStatement 对象的 executeUpdate() 方法执行添加操作
ps.executeUpdate();
```

第四步：在执行数据操作之后，需调用 PrepareStatement 对象和 Connection 对象的 close() 方法，从而及时释放 JDBC 所占用的数据库资源，关键代码如下。

```
if (resultSet != null) {resultSet.close();}
if (prepareStatement != null) {prepareStatement.close();}
if (connection != null) {conn.close();}
```

【实例】使用 JDBC 执行向 userinfo 表中添加一条数据的操作，实现代码如示例代码 5-2 所示。

示例代码 5-2：使用 JDBC 新增数据

```
package jdbc;
import java.sql.Connection;
import java.sql.DriverManager;
import java.sql.PreparedStatement;
import java.sql.SQLException;
public class JDBCDemo2 {
    public static void main(String[] args){
        try {
            Class.forName("com.MySQL.cj.jdbc.Driver");
            String url = "jdbc:MySQL://localhost:3306/ bookshop?"
              +"useUnicode=true&characterEncoding=utf8&serverTimezone=GMT%2B8";
            String user = "root";
            String password = "123456";
            Connection conn = DriverManager.getConnection(url, user, password);
            String sql = "INSERT INTO userinfo VALUES (NULL,?,?,?)";
            PreparedStatement ps = conn.prepareStatement(sql);
            ps.setString(1,"user1");
```

```
            ps.setString(2,"123456");
            ps.setString(3,"123@edu.com");
            int rows = ps.executeUpdate();
            System.out.println(" 数据添加成功,新增数据 "+rows+" 条 ");
            if (conn != null) {conn.close();}
            if (ps != null) {ps.close();}
        } catch (ClassNotFoundException e) {
            e.printStackTrace();
        } catch (SQLException e1){
            e1.printStackTrace();
        }
    }
}
```

代码运行后,控制台输出结果如图 5-4 所示。

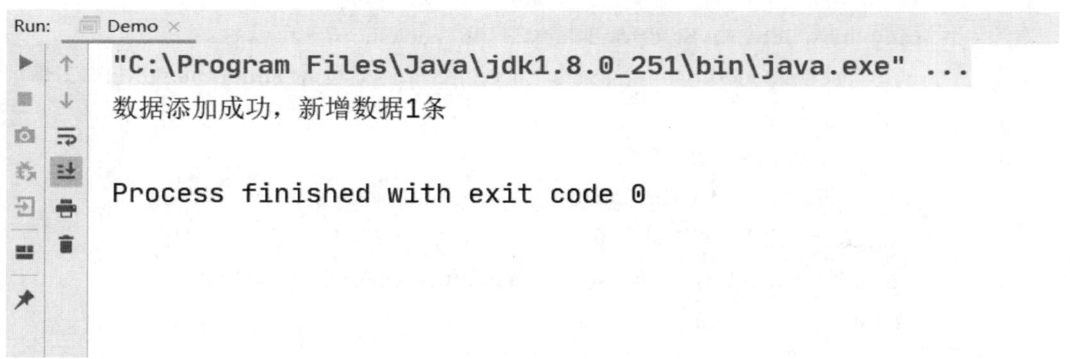

图 5-4　新增数据成功

查询 userinfo 表中数据的情况,验证新增数据是否添加成功,如图 5-5 所示高亮记录即为成功添加的用户信息。

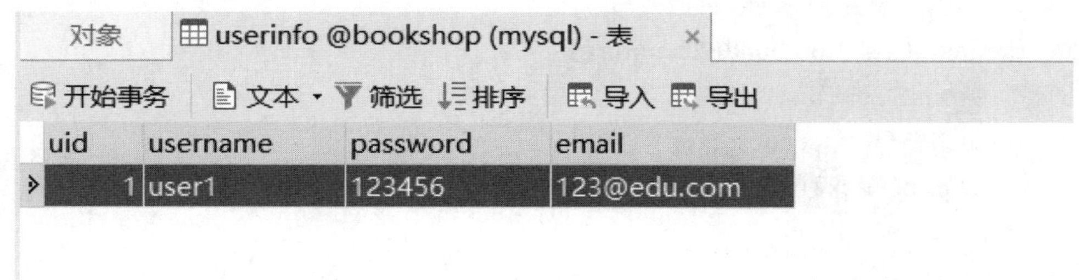

图 5-5　userinfo 表中数据信息

2. 修改、删除数据

使用 JDBC 修改、删除数据库中的数据，操作方法与添加数据相似，区别是修改、删除数据需要使用 UPDATE 和 DELETE 语句实现。

【实例】在上面实例的基础上，实现将 uid 为 1 的用户名（username）user1 改成 Tom，实现代码如示例代码 5-3 所示。

示例代码 5-3：修改数据

```
package jdbc;
import java.sql.Connection;
import java.sql.DriverManager;
import java.sql.PreparedStatement;
import java.sql.SQLException;
public class JDBCDemo3 {
    public static void main(String[] args){
        try {
            Class.forName("com.MySQL.cj.jdbc.Driver");
            String url = "jdbc:MySQL://localhost:3306/ bookshop?"
                    +"useUnicode=true&characterEncoding=utf8&serverTimezone=GMT%2B8";
            String user = "root";
            String password = "123456";
            Connection conn = DriverManager.getConnection(url, user, password);
            String sql = "UPDATE userinfo SET username = ? WHERE uid = ?";
            PreparedStatement ps = conn.prepareStatement(sql);
            ps.setString(1,"Tom");
            ps.setString(2,"1");
            int rows = ps.executeUpdate();
            System.out.println(" 数据修改成功，更新数据 "+rows+" 条 ");
            if (conn != null) {conn.close();}
            if (ps != null) {ps.close();}
        } catch (ClassNotFoundException e) {
            e.printStackTrace();
        } catch (SQLException e1){
            e1.printStackTrace();
        }
    }
}
```

代码运行后，控制台输出结果如图 5-6 所示。

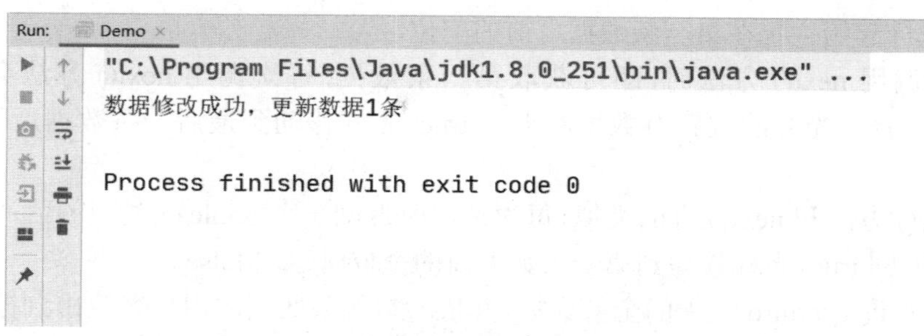

图 5-6　修改数据成功

查询 userinfo 表中数据的情况，验证修改数据是否成功，如图 5-7 所示，此时 username 为 Tom。

图 5-7　userinfo 表中数据信息

3. 查询数据

使用 JDBC 查询数据与新增、修改和删除的不同之处在于，查询操作调用 PreparedStatement 对象的 executeQuery() 方法执行查询 SQL 语句，执行完成后返回一个用来记录查询结果的集合对象 ResultSet，从数据库中查询到的数据内容和数据结构都存放在这个集合中。其结构如图 5-8 所示。

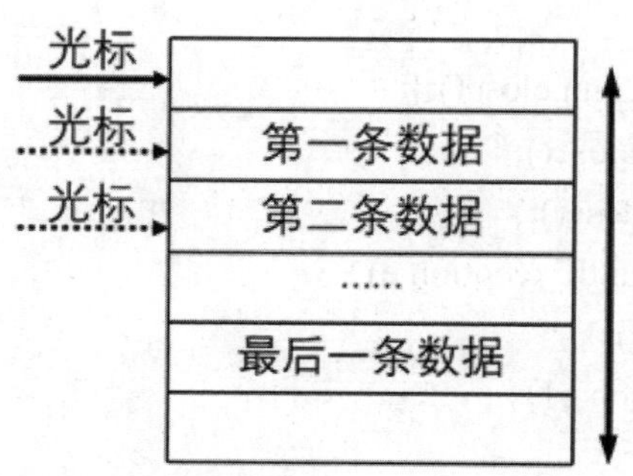

图 5-8　ResultSet 结构图

从图 5-8 中可以看出，在 ResultSet 集合中，通过移动光标依次获取查询到的数据，光标可以通过 next() 方法和 previous() 方法进行上下移动，如要获取某位置的数据，只需要把光标移动到该位置即可。ResultSet 结果集中，第一条数据之前和最后一条数据之后都有一个空白位置，光标起始位置在第一条数据之前的空位。

当首次调用 next() 方法时，即可获取第一条数据，每次调用 next() 方法会返回一个 boolean 类型的值，光标的位置有数据时返回 true，光标移动到最后一条数据后的空位时返回 false。

previous() 方法和 next() 方法类似，每次调用时返回一个 boolean 类型的值，当光标的位置有数据时返回 true，光标移动到第一条数据前的空位时返回 false。

【实例】查询 userinfo 表中的全部数据，并依次将每条数据的用户名输出到控制台上，实现代码如示例代码 5-4 所示。

示例代码 5-4：查询数据

```
package jdbc;
import java.sql.*;
public class JDBCDemo4 {
    public static void main(String[] args){
        try {
            Class.forName("com.MySQL.cj.jdbc.Driver");
            String url = "jdbc:MySQL://localhost:3306/ bookshop?"
             +"useUnicode=true&characterEncoding=utf8&serverTimezone=GMT%2B8";
            String user = "root";
            String password = "123456";
            Connection conn = DriverManager.getConnection(url, user, password);
            String sql = "SELECT * FROM userinfo";
            PreparedStatement ps = conn.prepareStatement(sql);
            ResultSet rs = ps.executeQuery(sql);
            while(rs.next()){
                System.out.println(" 用户名 :"+rs.getString("username")
                    +" 密码 :"+rs.getString("password")+" email:"+rs.getString("email"));
            }
            if (conn != null) {conn.close();}
            if (ps != null) {ps.close();}
            if (rs != null) {rs.close();}
        } catch (ClassNotFoundException e) {
            e.printStackTrace();
        } catch (SQLException e1){
            e1.printStackTrace();
        }
    }
}
```

提示：使用 while 条件循环遍历 ResultSet 对象，并通过 ResultSet 对象的 get () 方法获取当前行的数据，当遍历到末尾返回 false 时即可自动跳出循环。

代码运行后，控制台输出结果如图 5-9 所示。

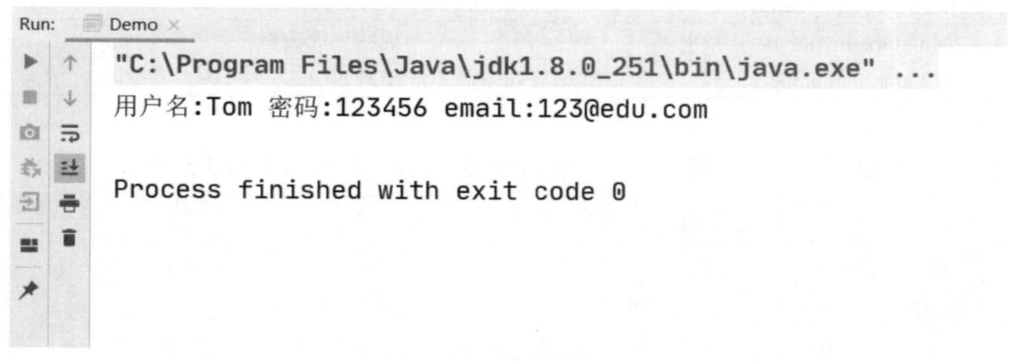

图 5-9　查询操作结果

技能点五　元数据

元数据是用来描述数据的数据，是从数据库中抽取出来的用于说明其特征、内容的结构化的数据（如列数、列名、列类型等）。在查询结果未知的情况下，数据库和结果集中的信息可以通过元数据对象获得。JDBC 提供了两个元数据对象，即 DatabaseMetaData 数据库元数据和 ResultSetMetaData 结果集元数据。

DatabaseMetaData 对象提供关于数据库的信息，包括以下几点。

1）数据库与用户、数据库标识符以及函数与存储过程。

2）数据库支持与不支持的功能。

3）数据库限制，如数据库中名称的最大长度。

4）架构、编目、表和列。

ResultSetMetaData 对象可以获得关于结果集对象的元数据信息，主要包含关于结果集中列的名称、类型为 NULL 或 NOT NULL、精度等信息。在数据库表中原有列之外，结果集还可以含有从表中派生来的信息（求和、计数）。通过调用 ResultSet 对象的 getMetaData() 方法可以获得 ResultSetMetaData 对象，相关代码如下。

```
ResultSet rs=stmt.executeQuery(sql);
ResultSetMetaData rsmd=rs.getMetaData();
```

得到 ResultSetMetaData 对象后，可以使用多种方法读取 ResultSet 对象的元数据。部分用于读结果集元数据的方法如表 5-6 所示。

表 5-6　ResultSetMetaData 常用方法

方法名	方法说明
getColumnCount()	返回结果集中的列数

续表

方法名	方法说明
getColumnName(int number)	返回 number 指定位置上的列的名称
getColumntype(int number)	返回列的类型
isNULLable(int number)	如果列被定义为 NOT NULL，返回 0；否则返回 1

【实例】获取并输出查询结果集中列名和列类型，代码如示例代码 5-5 所示。

示例代码 5-5：元数据应用

```
package jdbc;
import java.sql.*;
public class JDBCDemo5 {
  public static void main(String[] args){
    try {
      Class.forName("com.MySQL.cj.jdbc.Driver");
      String url = "jdbc:MySQL://localhost:3306/ bookshop?"
       +"useUnicode=true&characterEncoding=utf8&serverTimezone=GMT%2B8";
      String user = "root";
      String password = "123456";
      Connection conn = DriverManager.getConnection(url, user, password);
      String sql = "SELECT * FROM userinfo";
      PreparedStatement ps = conn.prepareStatement(sql);
      ResultSet rs = ps.executeQuery();
      ResultSetMetaData metaData = rs.getMetaData();
      int cols_len = metaData.getColumnCount();
      for(int i=0;i<cols_len;i++){
        System.out.println(" 列名："+metaData.getColumnName(i+1)
            +" 本列类型："+metaData.getColumnTypeName(i+1));}
      if (conn != null) {conn.close();}
      if (ps != null) {ps.close();}
      if (rs != null) {rs.close();}
    } catch (ClassNotFoundException e) {
      e.printStackTrace();
    } catch (SQLException e1){
      e1.printStackTrace();
    }
  }
}
```

代码运行后，控制台输出结果如图 5-10 所示。

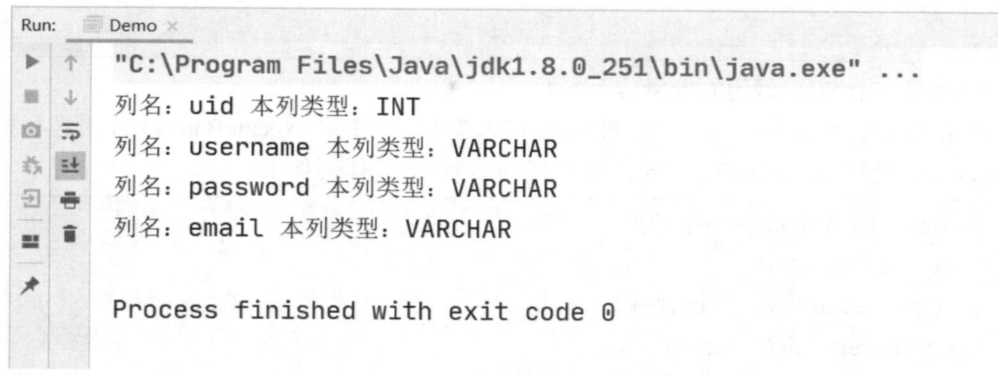

图 5-10　userinfo 表列名和列类型

技能点六　批处理

在 JDBC 开发中，操作数据库需要与数据库建立连接，然后将要执行的 SQL 语句传送到数据库服务器，最后关闭数据库连接。如果按照该流程执行多条 SQL 语句，就需要建立多个数据库连接，频繁的建立和关闭数据库连接需要消耗大量资源，使程序运行变慢。针对这一问题，JDBC 的批处理功能提供了很好的解决方案。

JDBC 中通过调用 batch 相关方法进行批处理操作，批处理的原理是建立一次连接，将批量 SQL 语句一次性发送到数据库中进行执行，从而解决多次与数据库连接所产生的速度瓶颈。JDBC 中进行批量处理的语句方法如下。

1）addBatch(String)：添加需要批量处理的 SQL 语句或是参数。

2）executeBatch()：执行批量处理语句。

3）clearBatch()：清除批量打包。

【实例】多条 SQL 语句的批处理。

使用 id、name、email 三个参数对 SQL 语句进行拼接，在 for 循环中使用 addBatch() 方法将需要批处理的 SQL 语句进行添加，添加完成后使用 executeBatch() 方法进行批处理，处理之后使用 clearBatch() 方法将积攒的参数列表进行清除，代码如下。

```
for (int i = 1; i < 5000; i++) {
    sql = "insert into person(id,name,email) values(" + i ",'name" + i + "','email" + i + "')";
    stmt.addBatch(sql);
    if ((i + 1) % 1000 == 0) {
// 批处理
stmt.executeBatch();
// 清除 stmt 中积攒的参数列表
    stmt.clearBatch();
    }
```

```
    }
```

【实例】一个 SQL 语句的批量传入参数。

使用 setInt() 和 setString() 方法对 SQL 传入参数，并使用 executeBatch() 方法进行批处理，处理之后使用 clearBatch() 方法清除积攒的参数列表，代码如下。

```
for(int i=1;i<100000;i++){
    pstmt.setInt(1, i);
    pstmt.setString(2, "name"+i);
    pstmt.setString(3, "email"+i);
    pstmt.addBatch();
    if((i+1)%1000==0){
// 批处理
    pstmt.executeBatch();
// 清空 pstmt 中积攒的 sql
    pstmt.clearBatch();
    }
}
```

在执行批处理过程中，需要注意使用 batch 和执行单 SQL 语句的不同点，具体有以下几点。

1）需要对同一个 PreparedStatement 反复设置参数并调用 addBatch()，就相当于给一个 SQL 加上了多组参数，相当于变成了“多行”SQL。

2）执行语句调用的不是 executeUpdate() 方法，而是 executeBatch() 方法。

3）batch 移交了多组参数，相应地，返回结果也是多个 int 值，因此返回类型是 int[]，循环 int[] 数组即可获取每组参数执行后影响的结果数量。

【实例】使用批处理方法向 userinfo 表中添加数据，代码如示例代码 5-6 所示。

示例代码 5-6：批量新增用户

```
package jdbc;
import java.sql.*;
public class JDBCDemo6 {
  public static void main(String[] args){
    try {
      Class.forName("com.MySQL.cj.jdbc.Driver");
      String url = "jdbc:MySQL://localhost:3306/ bookshop?"
        +"useUnicode=true&characterEncoding=utf8&serverTimezone=GMT%2B8";
      String user = "root";
      String password = "123456";
      Connection conn = DriverManager.getConnection(url, user, password);
      String sql = "INSERT INTO userinfo VALUES (NULL,?,?,?)";
      PreparedStatement ps = conn.prepareStatement(sql);
```

```
            for(int i=0;i<5;i++){
                ps.setString(1,"name_"+i);
                ps.setString(2,"password_"+i);
                ps.setString(3,"email_"+i+"@edu.com");
                // 执行批处理
                        ps.addBatch();
            }
            int[] rows = ps.executeBatch();
            System.out.println(" 本次共插入数据 "+rows.length+" 条 ");
                if (conn != null) {conn.close();}
            if (ps != null) {ps.close();}
            if (conn != null) {conn.close();}
        } catch (ClassNotFoundException e) {
            e.printStackTrace();
        } catch (SQLException e1){
            e1.printStackTrace();
        }
    }
}
```

代码运行后，控制台输出结果如图 5-11 所示。

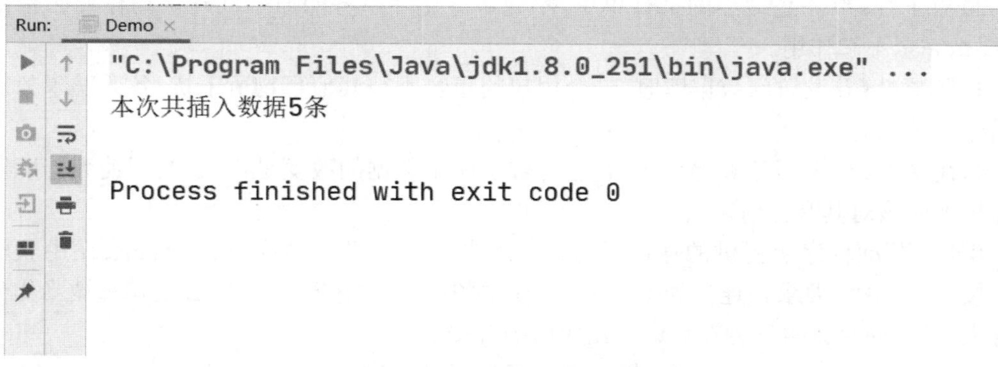

图 5-11　批量新增用户成功

查询 userinfo 表中数据的情况，验证批量新增数据是否成功，如图 5-12 所示，批量增加 5 条数据。

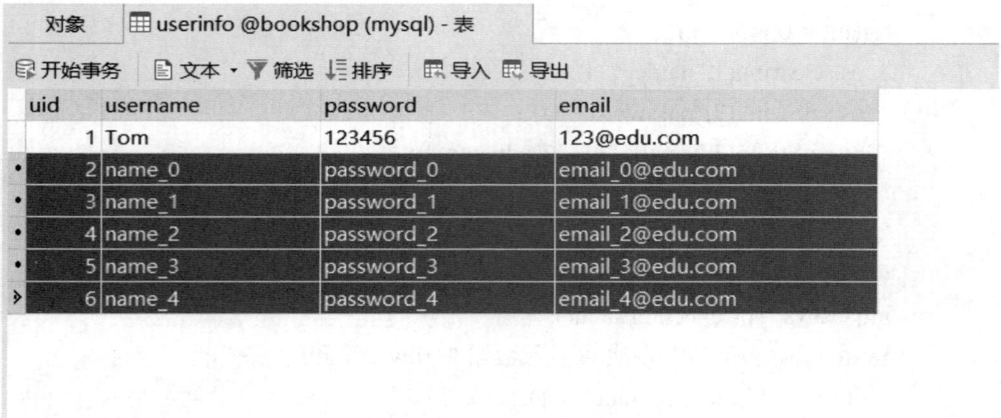

对象 userinfo @bookshop (mysql) - 表

开始事务 文本 · 筛选 排序 导入 导出

uid	username	password	email
1	Tom	123456	123@edu.com
2	name_0	password_0	email_0@edu.com
3	name_1	password_1	email_1@edu.com
4	name_2	password_2	email_2@edu.com
5	name_3	password_3	email_3@edu.com
6	name_4	password_4	email_4@edu.com

图 5-12 userinfo 表中数据信息

技能点七 JDBC 事务

数据库事务是指由若干个 SQL 语句构成的一个操作序列，类似于 Java 的 synchronized 同步功能。数据库系统保证在一个事务中的所有 SQL 要么全部执行成功，要么全部不执行，数据库事务具有四个特性，具体内容如下。

1）一致性：执行事务前后，数据保持一致。

2）原子性：事务是最小的执行单位，不允许分割。事务的原子性确保动作要么全部完成，要么完全不起作用。

3）隔离性：并发访问数据库时，一个用户的事物不被其他事物所干扰，各并发事务之间数据库是独立的。

4）耐久性：一个事务被提交之后，它对数据库中数据的改变是持久的，即使数据库发生故障也不应该对其有任何影响。

JDBC 驱动程序所提供的连接类将提供事务控制。当代码从 DriverManager 中获取一条连接时，JDBC 要求该连接处于自动提交模式中。这意味着每条 SQL 语句都被当作一个事务来对待，而且该事务会在该语句结束时被提交。

JDBC 通过 Connection 对象的 setAutocommit(boolean) 方法决定是否启用自动提交。当自动提交模式被设置成 false 时，事务管理现在必须由代码来明确控制。而代码通过在适当的时候调用连接对象的适当方法完成它的任务。发送给数据库的 SQL 语句仍将得到执行，但该事务在语句执行完毕时不被提交。该事务将等到代码调用了连接对象的 commit() 方法时才会提交。另一种情况是如果代码调用连接对象的 rollback() 方法，该事务可以回退。

【实例】利用事务完成同时向 userinfo 表和 books 表中插入数据，代码如示例代码 5-7 所示。

示例代码 5-7：使用事务进行数据新增

```
package jdbc;
import java.sql.*;
public class JDBCDemo7 {
    public static void main(String[] args) {
        Connection conn = null;
        PreparedStatement ps = null;
        try{
            Class.forName("com.MySQL.cj.jdbc.Driver");
            String url = "jdbc:MySQL://localhost:3306/ bookshop?"
             +"useUnicode=true&characterEncoding=utf8&serverTimezone=GMT%2B8";
            String user = "root";
            String password = "123456";
            conn = DriverManager.getConnection(url, user, password);
            conn.setAutoCommit(false);    // 关闭自动提交 , 开启事务
            String sql1 = "INSERT INTO userinfo VALUES (NULL,?,?,?)";
            ps = conn.prepareStatement(sql1);
            ps.setString(1,"Tony");
            ps.setString(2,"111222");
            ps.setString(3,"tony@edu.com");
            ps.executeUpdate();
            String sql2 = "INSERT INTO books VALUES(NULL,?,?,?,?)";
            ps = conn.prepareStatement(sql2);
            ps.setString(1,"Algorithm");
            ps.setString(2,"70");
            ps.setString(3,"c:\\picture\\Algorithm.jpg");
            ps.setString(4,"10");
            ps.executeUpdate();
            conn.commit(); // 提交事务
            System.out.println(" 事务执行成功 ");
            if (ps != null) {ps.close();}
            if (conn != null) {conn.close();}
        }catch (ClassNotFoundException e1){
            e1.printStackTrace();
        }catch (SQLException e2){
            try {
```

```
                System.out.println(" 事务执行出错，执行回滚操作 ");
                e2.printStackTrace();
                conn.rollback(); // 如 SQL 操作出错，执行回滚操作
            }catch (SQLException e3){
                System.out.println(" 回滚操作出错 ");
                e2.printStackTrace();
            }
        }
    }
}
```

代码运行后，控制台输出结果如图 5-13 所示。

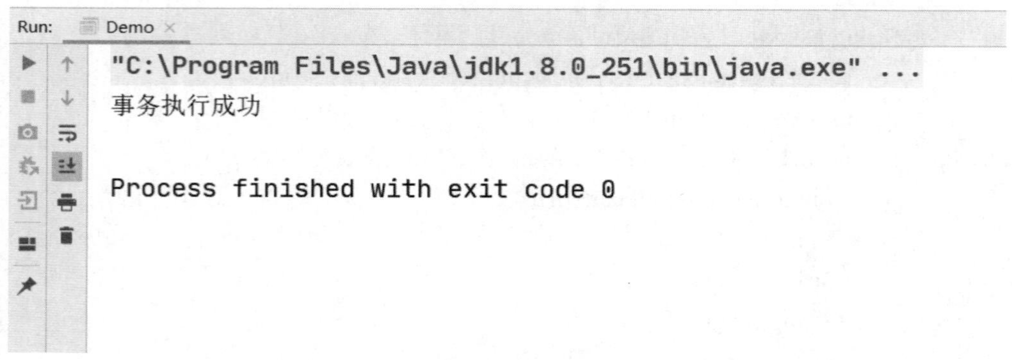

图 5-13　事务执行成功

查询 userinfo 表和 books 表中数据的情况，验证事务中添加用户信息以及添加书籍信息的操作是否成功，如图 5-14 和图 5-15 所示，高亮记录即为成功新增的用户信息和书籍信息。

对象　userinfo @bookshop (mysql) - 表　books @bookshop (mysql) - 表

开始事务　文本 · 筛选　排序　导入　导出

uid	username	password	email
1	Tom	123456	123@edu.com
2	name_0	password_0	email_0@edu.com
3	name_1	password_1	email_1@edu.com
4	name_2	password_2	email_2@edu.com
5	name_3	password_3	email_3@edu.com
6	name_4	password_4	email_4@edu.com
7	Tony	111222	tony@edu.com

图 5-14　userinfo 表中数据信息

图 5-15　books 表中数据信息

技能点八　Properties 类

Properties（Java.util.Properties）是 Java 语言的配置文件所使用的类，用于读取 Java 的配置文件。配置文件在不同的系统中设置有所不同，对于不同的用户会存在不同的设置。对于这种特殊的文件，文件中的很多变量需要改变，为了能让用户脱离系统本身去修改相关的变量设置以应对自己的需求，需要将系统配置相关的内容整合在一个文件中以方便修改。例如在 Java 中，其配置文件常为 .properties 文件，以 key-value 的形式进行参数配置。

读取配置文件需要用到 Properties 类，该类继承于 Hashtable，表示一个持久的属性集，属性列表中每个键及其对应值都是字符串类型。该类的常用方法如表 5-7 所示。

表 5-7　Properties 常用方法

方法名	方法说明
void load(InputStream streamIn) throws IOException	从输入流中读取属性列表（键值对）
void list(PrintWriter streamOut)	将属性列表输出到指定的输出流
void list(PrintStream streamOut)	将属性列表输出到指定的输出流
String getProperty(String key)	用指定的键在此属性列表中搜索属性
Object setProperty(String key, String value)	向属性列表中写入属性
Enumeration propertyNames()	返回属性列表中所有键的枚举

【实例】创建一个配置文件 file.properties，并读取其中的内容。

创建一个 file.properties 文件，内容为两组键值对 age=23、address=tianjin，如示例代码 5-8 所示。

示例代码 5-8：file. properties

```
age = 23
address = tianjin
```

创建 myProperties 类，运用 load() 方法读取 file 配置文件中的数据，使用 propertyNames() 方法获取所有键值对，并使用 getProperty() 方法获取具体数据，如示例代码 5-9

所示。

示例代码 5-9：myProperties.java

```
package com.xt.util;
import java.io.FileInputStream;
import java.io.InputStream;
import java.util.Enumeration;
import java.util.Properties;

class myProperties {
    public static void main(String[] args) throws Exception {
        Properties pps = new Properties();
            InputStream inputStream = myProperties.class.getClassLoader().getResourceAs-
Stream("file.properties");
// 寻找根目录下的 file 配置文件
        pps.load(inputStream);
// 从输入流中读取属性列表
        Enumeration fileName = pps.propertyNames();
        // 获取属性列表中所有键的枚举
      while (fileName.hasMoreElements()) {
          String strKey = (String) fileName.nextElement();// 获取 key
          String strValue = pps.getProperty(strKey);// 根据 key 获取 value
           System.out.println(strKey + "," + strValue);// 输出在控制平台中
        }
    }
}
```

运行后会在控制台显示 file 配置文件中的数据，效果如图 5-16 所示。

```
"C:\Program Files\Java\jdk1.8.0_25\bin\java.exe" ...
address,tianjin
age,23

Process finished with exit code 0
```

图 5-16 file 配置文件中的数据

实现网上书城项目数据库的连接，封装对应的增加、删除、修改和查询的操作，具体步骤

如下。

第一步：创建网上书城数据库表。

根据第一章中提供的数据结构，编写 SQL 语句，创建数据库表。其中，用户信息表（userinfo）包含用户名、密码和邮箱，SQL 语句如示例代码 5-10 所示。

示例代码 5-10：创建 userinfo 表

```
CREATE TABLE userinfo (
UID INT NOT NULL AUTO_INCREMENT,
username VARCHAR(50),
password VARCHAR(50),
email VARCHAR(50),
PRIMARY KEY (UID));
```

数据库 userinfo 表结构如图 5-17 所示。

对象　userinfo @bookshop (mysql) - 表

保存　添加字段　插入字段　删除字段　主键

字段　索引　外键　触发器　选项　注释　SQL 预览

名	类型	长度	小数点	不是 null	键
UID	int	0	0	☑	1
username	varchar	50	0	☐	
password	varchar	50	0	☐	
email	varchar	50	0	☐	

图 5-17　userinfo 表结构图

图书信息表 books 包含图书的名称、价格、封皮等信息，创建图书信息表 SQL 语句如示例代码 5-11 所示。

示例代码 5-11：创建 books 表

```
CREATE TABLE books (
BID INT NOT NULL AUTO_INCREMENT,
bookname VARCHAR(100),
b_price VARCHAR(10),
image VARCHAR(200),
stock INT,
PRIMARY KEY (BID));
```

创建完毕的 books 表结构如图 5-18 所示。

对象 | books @bookshop (mysql) - 表

保存 添加字段 插入字段 删除字段 主键

字段 索引 外键 触发器 选项 注释 SQL 预览

名	类型	长度	小数点	不是 null	键
BID	int	0	0	☑	1
BOOKNAME	varchar	100	0	☐	
B_PRICE	varchar	10	0	☐	
IMAGE	varchar	200	0	☐	
STOCK	int	0	0	☐	

图 5-18 books 表结构图

第二步：将 JDBC 流程封装成新增方法 doInsert()、修改删除方法 doUD() 和查询方法 doSelect()，并将获取 JDBC 连接和关闭 JDBC 资源的过程分别封装成 openconnection() 与 closeResource() 方法。在项目中创建 dao 包存放数据库操作相关的类，在包中创建 BaseDao.java 类编写封装代码。项目结构如图 5-19 所示。

图 5-19 创建 BaseDao 类

第三步：创建读取配置文件的工具类。创建 util 包用于存放工具类，包中创建 ConfigManager 类封装读取配置文件的方法，并在 com.xt 包下创建 database.properties 文件存储连接用到的参数，如图 5-20 所示。

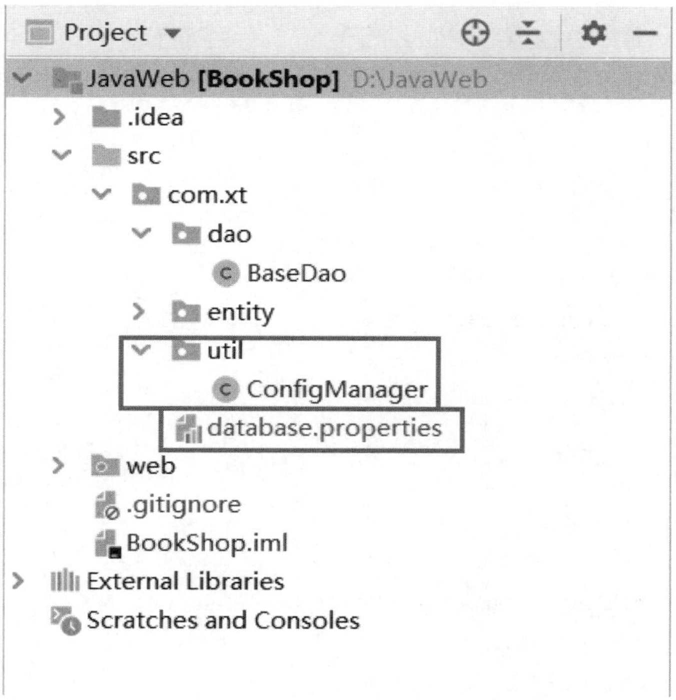

图 5-20　配置文件相关创建

第四步：编写 ConfigManager 类，该类实例以单例模式进行创建，这样可以减少内存的开销，避免对资源的多重占用（如 IO 操作），类代码如示例代码 5-12 所示。

示例代码 5-12：ConfigManager.java

```
package com.xt.util;
import java.io.IOException;
import java.io.InputStream;
import java.util.Properties;
/**
 * 读取 properties 配置文件工具类
 */
public class ConfigManager {
    private static ConfigManager configManager;
    // 用于读取配置文件
    private static Properties properties;
    /**
* 将 properties 配置文件读取到 Properties 文件中
*/
    private ConfigManager(){
    String configFile="database.properties";
```

```
        properties=new Properties();
        InputStream  in=ConfigManager.class.getClassLoader().getResourceAsStream(config-
File);
        try {
        properties.load(in);
        in.close();
        } catch (IOException e) {
        // TODO Auto-generated catch block
        e.printStackTrace();
        }
        }
        /**
* 获取类对象的单例对象
*/
        public static ConfigManager getInstance(){
        if(configManager==null){
        configManager=new ConfigManager();
        }
        return configManager;
        }
        /**
* 读取配置文件中属性值
*/
        public String getString(String key){
        return properties.getProperty(key);
        }
    }
```

在 database.properties 文件中以键值对的形式保存 JDBC 的驱动类全类名、连接 url、连接用户名和密码,代码如示例代码 5-13 所示。

示例代码 5-13：database properties

```
jdbc.driver_class=com.MySQL.cj.jdbc.Driver
jdbc.connection.url=jdbc:MySQL://localhost:3306/bookshop?useUnicode=true&charac-
terEncoding=utf8&zeroDateTimeBehavior=convertToNull&useSSL=true&serverTimezone=G-
MT%2B8
jdbc.connection.username=root
jdbc.connection.password=123456
```

第五步: 声明 JDBC 对象。在 BaseDao 封装类中声明 JDBC 执行流程中的 3 个重要对象 Connection、PreparedStatement 和 ResultSet，声明代码如下。

```
protected Connection conn = null;
protected PreparedStatement ps = null;
protected ResultSet rs = null;
```

第六步：封装 openConnection() 方法。openConnection() 方法用来实现注册 JDBC 驱动，创建 Connection 连接对象实例，并将创建的 Connection 实例赋值给类中声明的 Connection 对象，以供类中其他封装方法调用，代码如示例代码 5-14 所示。

示例代码 5-14：获取连接对象方法 openConnection()

```
public void openConnection(){
  // 读取 properties 文件中的连接参数
String driver=ConfigManager.getInstance().getString("jdbc.driver_class");
String url=ConfigManager.getInstance().getString("jdbc.connection.url");
String username=ConfigManager.getInstance().getString("jdbc.connection.username");
String password=ConfigManager.getInstance().getString("jdbc.connection.password");
try {
    Class.forName("com.MySQL.cj.jdbc.Driver");
    conn = DriverManager.getConnection(url, username, password);
} catch (ClassNotFoundException e) {
    e.printStackTrace();
} catch (SQLException e) {

    e.printStackTrace();
    }
}
```

第七步：封装 closeResource() 方法。closeResource() 方法用来实现关闭存在的 Connection 对象、Statement 对象和 ResultSet 对象，代码如示例代码 5-15 所示。

示例代码 5-15：关闭资源方法 closeResource ()

```
public boolean closeResource(){
    try {

    if(rs != null)

    rs.close();

    if(ps != null)
```

```
        ps.close();

        if(conn != null)

        conn.close();
        } catch (SQLException e) {


        e.printStackTrace();

        return false;
        }
        return true;
    }
```

第八步：封装 doInsert() 方法。doInsert() 方法用来实现通过传入 SQL 语句执行新增方法，并返回新增数据的自增主键。doInsert() 方法如示例代码 5-16 所示。

示例代码 5-16：新增方法封装

```
    public int doInsert(String sql, Object[] params){
        int ID = 0;
        try {
        openConnection();
        ps = conn.prepareStatement(sql,Statement.RETURN_GENERATED_KEYS);
        for (int i = 0; i < params.length; i++) {
        ps.setObject(i + 1, params[i]);
        }
        ps.executeUpdate();
        ResultSet rs = ps.getGeneratedKeys();
        if(rs.next()){
        ID = rs.getInt(1);
        }
        } catch (Exception e) {
        e.printStackTrace();
        } finally {
        closeResource();
        }
        return ID;
    }
```

第九步：封装 doUD() 方法。doUD() 方法用来实现通过传入 SQL 语句执行修改和删除方法，并返回修改和删除语句影响的数据条目数。该方法需传入 2 个参数，分别是 String 类型含有占位符的 Insert 语句以及为占位符赋值的 Object 数组，代码如示例代码 5-17 所示。

示例代码 5-17：修改、删除方法封装

```
public int doUD(String sql, Object[] params) {
    int rows = 0;
    try {
    openConnection();
    ps = conn.prepareStatement(sql);
    for (int i = 0; i < params.length; i++) {
    ps.setObject(i + 1, params[i]);
    }
    rows = ps.executeUpdate();
    } catch (Exception e) {
    e.printStackTrace();
    } finally {
    closeResource();
    }
    return rows;
}
```

第十步：封装 doSelect() 方法。doSelect() 方法用来实现通过传入 SQL 语句执行查询方法，并返回封装了查询结果的 List 对象，List 集合对象中保存了以 Map 集合形式存储的单行数据。该方法需传入 2 个参数，分别是 String 类型含有占位符的 Insert 语句以及为占位符赋值的 Object 数组，代码如示例代码 5-18 所示。

示例代码 5-18：查询方法封装

```
public List doSelect(String sql, Object[] params) {
    List<Map> list = new ArrayList<Map>();
    try {
    openConnection();
    ps = conn.prepareStatement(sql);
    for (int i = 0; i < params.length; i++) {
    ps.setObject(i + 1, params[i]);
    }
    rs = ps.executeQuery();
    ResultSetMetaData metaData = rs.getMetaData();
    int cols_len = metaData.getColumnCount();
```

```
        while (rs.next()) {
        Map map = new HashMap();
        for (int i = 0; i < cols_len; i++) {
        String cols_name = metaData.getColumnName(i + 1);
        Object cols_value = rs.getObject(cols_name);
        if (cols_value == null) {
        cols_value = "";
        }
        map.put(cols_name, cols_value);
        }
        list.add(map);
        }
        } catch (Exception e) {
        e.printStackTrace();
        } finally {
        closeResource();
        }
        return list;
    }
```

第十一步：验证封装方法。使用封装后的方法执行查询，并在控制台中输出 userinfo 表中的数据，代码如示例代码 5-19 所示。

示例代码 5-19：验证封装方法

```
public static void main(String[] args) {
BaseDao baseDao = new BaseDao();
String sql = "Select * FROM userinfo";
List<HashMap> list = baseDao.doSelect(sql,new Object[]{});
    for(HashMap map:list){
    System.out.println(map);
}}
```

代码运行后，控制台输出结果如图 5-21 所示。

```
"C:\Program Files\Java\jdk1.8.0_251\bin\java.exe" ...
{uid=1, password=123456, email=123@edu.com, username=Tom}
{uid=2, password=password_0, email=email_0@edu.com, username=name_0}
{uid=3, password=password_1, email=email_1@edu.com, username=name_1}
{uid=4, password=password_2, email=email_2@edu.com, username=name_2}
{uid=5, password=password_3, email=email_3@edu.com, username=name_3}
{uid=6, password=password_4, email=email_4@edu.com, username=name_4}
{uid=7, password=111222, email=tony@edu.com, username=Tony}

Process finished with exit code 0
```

图 5-21 封装方法执行查询操作结果

本次任务通过对 JDBC 基本概念、连接对象的学习，重点熟悉 JDBC 操作数据库的方法，了解元数据、批处理以及事务的知识，学会如何创建配置文件设置数据库的连接，为以后网上书城项目的整体制作奠定基础。

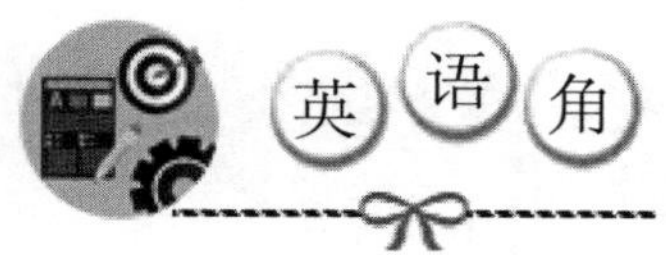

connection	连接	statement	声明
protocol	协议	property	属性
execute	执行	batch	批量

1. 选择题

1）使用 Connection 的（　　）方法可以建立一个 PreparedStatement 接口。

A. createPrepareStatement ()　　B. prepareStatement ()

C. createStatement ()　　D. Statement ()

2）下面关于 Preparedstatement 的说法错误的是（　　）。

A. PreparedStatement 继承了 Statement

B. PreparedStatement 可以有效防止 SQL 注入

C. PreparedStatement 不能用于批量更新的操作

D. PreparedStatement 可以存储预编译的 Statement，从而提升执行效率

3）下面描述中错误的是（　　）。

A. Statement 的 executeQuery() 方法会返回一个结果集

B. Statement 的 executeUpdate() 方法会返回是否更新成功的 boolean 值

C. ResultSet 中的 getString() 方法可以获得一个对应于数据库中 char 类型的值

D. Resultset 中的 next() 方法会使结果集中的下一行成为当前行

4）下面有关 ResultSet 的说法错误的是（　　）。

A. 如果 JDBC 执行查询语句没有查询到数据，那么 ResultSet 将会是 null 值

B. 如果使用 next() 方法将 ResultSet 光标移动到末尾，那么 next() 方法会返回 false

C. 如果 Connection 对象关闭，那么 ResultSet 也无法使用

D. 如果一个事物没有提交，那么 ResultSet 中看不到事物过程中的临时数据

5）在 JDBC 中使用事务，回滚事务的方法是（　　）。

A. connection 的 commit()　　　　B. connection 的 setAutoCommit()

C. connection 的 rollback()　　　　D. connection 的 close ()

2. 简答题

简述使用 JDBC 查询数据库中内容，并在控制平台中输出结果的全部操作流程。

第六章　网上书城项目登录注册功能

通过学习 Servlet 相关知识，并结合之前所讲的 JSP 知识，了解并掌握 Servlet 的原理、概念以及基本方法，实现网上书城项目的登录和注册功能。在任务实现过程中：

- 了解 Servlet 基本概念、原理、特点、生命周期；
- 掌握 Servlet 配置方法；
- 掌握 Servlet 读取表单、Session 的方法；
- 掌握过滤器、监听器的使用。

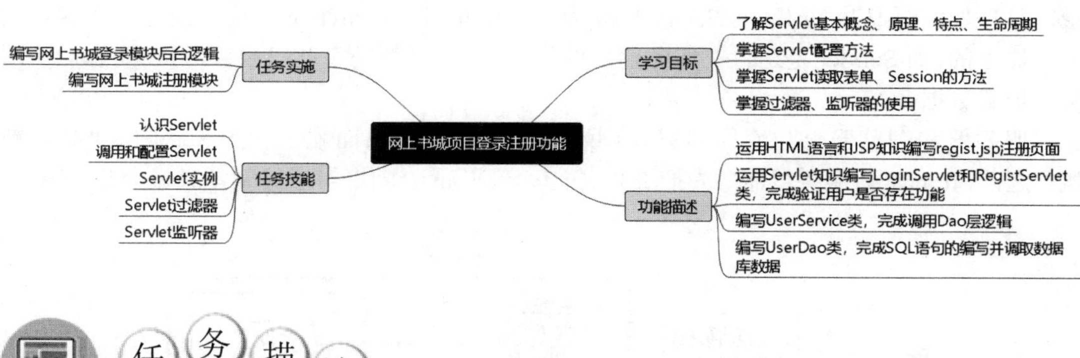

【情境导入】

在构建一个完整的项目时，需要对用户的请求信息进行应答。例如，用户在前端填写表单信息，需要将信息传递至后端进行处理，并存入数据库，再获取信息并传递到前端页面。

完成这一过程，需要交互式浏览和修改数据，并生成动态 Web 内容，此时就需要使用 Servlet 完成。

【功能描述】

● 运用 HTML 语言和 JSP 知识编写 regist.jsp 注册页面；

● 运用 Servlet 知识编写 LoginServlet 和 RegistServlet 类，完成验证用户是否存在等功能；

● 编写 UserService 类，完成调用 Dao 层逻辑；

● 编写 UserDao 类，完成 SQL 语句的编写并调取数据库数据。

技能点一 认识 Servlet

1.Servlet 概念

Java Servlet 是运行在 Web 服务器或应用服务器上的程序，它一般作为来自 Web 浏览器或其他 HTTP 客户端的请求和 HTTP 服务器上的数据库以及应用程序之间的中间层。

使用 Servlet 可以收集来自网页表单的用户输入，呈现来自数据库或者其他数据源的记录，还可以动态创建网页。Servlet 与 Java 应用程序相对应，Java 应用程序是运行在客户端浏览器上的，而 Servlet 是运行在服务器端的程序，但两者都是字节码对象，可以动态地从网络上加载数据。

服务器和浏览器交互流程如图 6-1 所示，每一个 JSP 页面就是一个 Servlet。JSP 在被执行之前，Tomcat 等服务器需要先把 JSP 页面编译成 Java 源代码，最终编译成一个 Servlet。

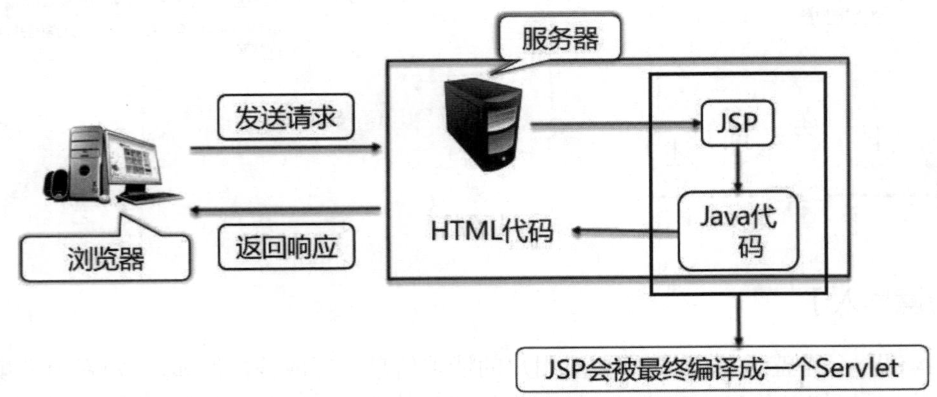

图 6-1 服务器和浏览器交互流程

通过图 6-1 可知，JSP 的所有功能都可以使用 Servlet 完成，Servlet 具体功能如下：

1）创建并返回基于客户请求的动态 HTML 页面；

2）创建可嵌入到现有的 HTML 页面中的部分 HTML 页面（HTML 片段）；

3）与其他服务器资源（如数据库、JavaBean 等）进行通信；

4）接收多个客户机的输入，并将结果广播到多个客户机上，例如 Servlet 可以实现支持多个参与者的游戏服务器；

5）根据客户请求采用特定 MIME（Multipurpose Internet Mail Extensions）类型对数据过滤，例如进行图像格式的转换。

2. Servlet 的特点

Servlet 程序在服务器端运行，能够动态生成 Web 页面。与早期的 Web 服务器和许多类似技术相比，Servlet 具有更高的效率，更容易使用，更强大的功能，更好的可移植性，其特点包含以下几个方面。

1）高效率：早期的 Web 服务器，每次都要请求一个新的进程；在 Servlet 中，每个请求都由一个轻量级的 Java 线程处理，可以缓冲以前的计算结果，提高数据库的效率。

2）更方便：Servlet 提供了大量的实用工具程序，使得表单数据的处理和 Cookie、Session 的追踪处理等更方便快捷。

3）功能强大：Servlet 能够直接与 Web 服务器进行交互。

4）可移植性好：均使用 Java 语言编写，拥有完整的标准，在编写 Servlet 时无须任何实质上的改动即可移植到其余的平台或是 Web 服务器。

3.Servlet API

Servlet API 又称为 Java Servlet 应用程序接口，包含了很多 Servlet 中重要的接口和类，具体如表 6-1 所示。

表 6-1　Servlet 重要接口和类

名称	说明
Servlet 接口	Java Servlet 的基本接口，定义了 Servlet 必须实现的方法
GenericServlet 接口	继承自 Servlet 接口，属于通用的、不依赖于协议的 Servlet
HttpServletResquest 接口	继承自 ServletResquset 接口，用于获取请求数据
HttpServletResponse 接口	继承自 ServletResponse 接口，用于返回响应数据
HttpServlet 类	继承自 GenericServlet 类，是在其基础上扩展了 HTTP 协议的 Servlet

Servlet API 的优势表现在以下几个方面。

1）Servlet 可以和其他资源交互，以生成返回给客户端的响应内容，也可以根据用户需要保存“请求—响应”过程中的信息。

2）采用 Servlet 技术，服务器可以完全授权对本地资源的访问，并且 Servlet 自身将会控制外部用户的访问数量及访问性质。

3）Servlet 可被链接。

4.Servlet 生命周期

Servlet 生命周期可被定义为从创建直到销毁的整个过程，主要分为 4 个阶段，分别是加载类和实例化（为对象分配空间）、初始化（为对象的属性赋值）、请求处理（服务阶段）、销毁。在 Servlet 生命周期中，会用到 init()、service() 等方法，具体如表 6-2 所示。

表 6-2 Servlet 重点方法

方法	说明
init()	在 Servlet 的生命周期中，仅执行一次 init() 方法。它是在服务器装入 Servlet 时执行的，可以重启服务器，在启动服务器或客户机首次访问 Servlet 时装入 Servlet
service()	service() 方法是 Servlet 的核心。每当一个客户请求一个 HttpServlet 对象时，该对象的 service() 方法就要被调用，传递给这个方法一个“请求”（ServletRequest）对象和一个“响应”（ServletResponse）对象作为参数。在 HttpServlet 中已存在 service() 方法
doGet()	当一个客户通过 HTML 表单发出一个 GET 请求或通过 URL 直接传递参数时，doGet() 方法被调用。与 GET 请求相关的参数添加到 URL 的后面，并与这个请求一起发送。当不会修改服务器端的数据时，应该使用 doGet() 方法
doPost()	当一个客户通过 HTML 表单发出一个 POST 请求时，doPost() 方法被调用。与 POST 请求相关的参数作为一个单独的请求从浏览器发送到服务器。当需要修改服务器端的数据时，应该使用 doPost() 方法
destroy()	destroy() 方法仅执行一次，即在服务器停止且卸装 Servlet 时执行

Servlet 生命周期如图 6-2 所示。

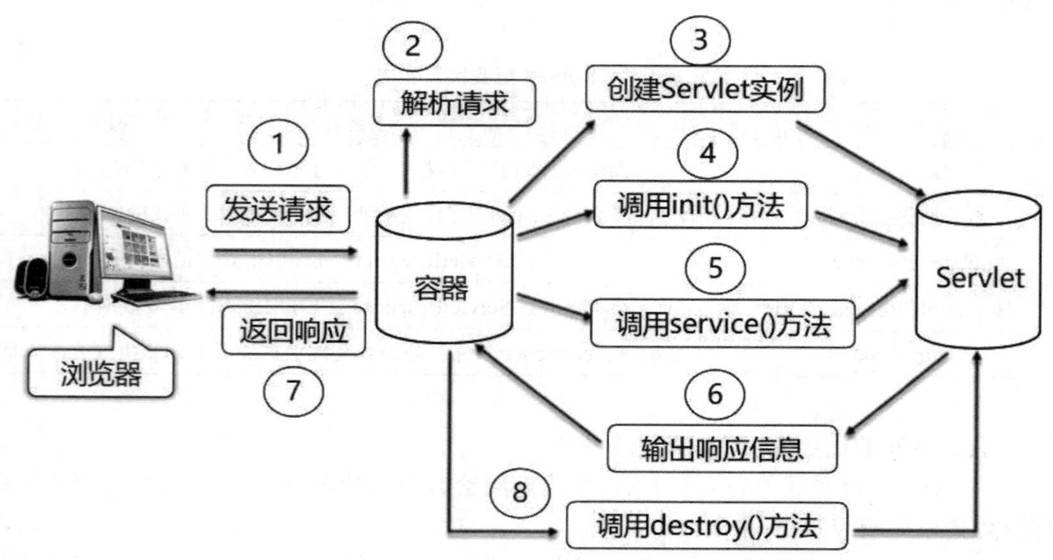

图 6-2 Servlet 生命周期

（1）加载和实例化（为对象分配空间）

当 Servlet 容器启动或客户端发送一个请求时，Servlet 容器会查找内存中是否存在该

Servlet 实例，如果有对应的实例，直接读取实例响应请求；如果不存在对应实例，就创建一个 Servlet 实例。

（2）初始化（为对象的属性赋值）

实例化后，Servlet 容器将调用 Servlet 的 init() 方法进行初始化（一些准备工作或资源预加载工作）。

（3）请求处理（服务阶段）

当接收到客户端请求时，调用 service() 方法处理客户端请求，HttpServlet 的 service() 方法会根据不同的请求调用（do Get() 或 do Post）方法。

（4）销毁

当 Servlet 容器关闭时，Servlet 实例也随之销毁。除此之外，Servlet 容器会调用 Servlet 的 destroy() 方法判断该 Servlet 是否应当被释放（或回收资源）。

技能点二　调用和配置 Servlet

1.Servlet 常用类与接口的层次关系

Java API 提供了 javax.servlet 和 javax.servlet.http 包，为编写 Servlet 提供接口和类。所有 Servlet 都必须实现 Servlet 接口，该接口定义了 Servlet 的生命周期方法，当实现一个通用的服务时，可以使用或继承 GenericServlet 类。编写 Servlet 时用到的主要 Servlet 层次结构如图 6-3 所示。

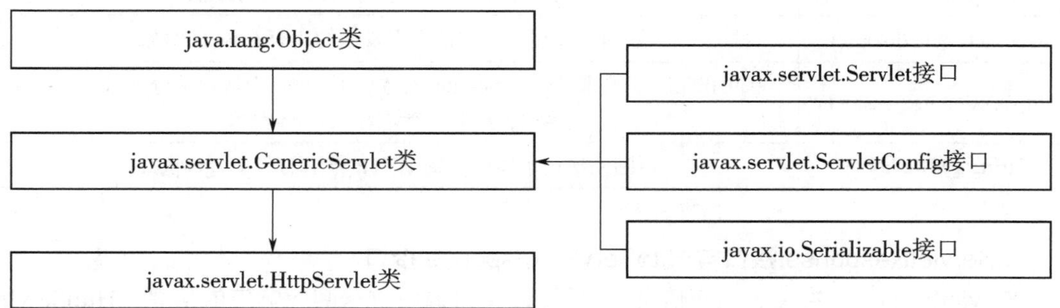

图 6-3　Servlet 常用类与接口的层次关系

2. ServletRequest 接口与 HttpServletRequest 接口

（1）ServletRequest 接口

当客户端发送请求时，由 Servlet 容器创建 Servlet 对象用于封装客户的请求信息，这个对象将被容器作为 service() 方法的参数之一传递给 Servlet，Servlet 能够利用 ServletRequest 对象获取客户端的请求数据。ServletRequest 接口常用方法见表 6-3。

表 6-3 ServletRequest 接口常用方法

类 / 接口名	说明
getContentLength()	返回请求正文的长度。如果请求正文的长度未知，则返回 -1
getContentType()	获得请求正文的 MIME 类型。如果请求正文的类型未知，则返回 null
getInputStram()	返回用于读取请求正文的输入流
getLocalAddr()	返回服务器端的 IP 地址
getLocalName()	返回服务器端的主机名
getParameter（String name）	根据给定的请求参数名，返回来自客户请求的匹配的请求参数值
getProtocal()	返回客户端和服务器端通信所用的协议的名称及版本号
getReader()	返回用户读取字符串形式的请求正文的 BufferedReader 对象
getRemoteAddr()	返回客户端的 IP 地址
getRemoteHost()	返回客户端的主机名

（2）HttpServletRequest 接口

HttpServletRequest 继承自 ServletRequest 接口，除继承了一些方法之外，还增加了一些用于读取请求信息的方法。HttpServletRequest 接口常用方法见表 6-4。

表 6-4 HttpServletRequest 接口常用方法

方法	说明
String getContextPath()	返回请求 URI 中表示请求上下文的路径
Cookie[] getCookies()	返回客户端在此次请求中发送的所有 Cookie 对象
HttpSession getSession()	返回和此次请求相关联的 Session 对象，如果没有给客户端分配 Session 对象，则创建一个新的 Session 对象
String getMethod()	返回此次请求所使用的 HTTP 方法的名字（get/post）

3.ServletResponse 接口与 HttpServletResponse 接口

ServletResponse 封装了响应信息，可通过该接口调用方法回应响应的信息。HttpServletResponse 是 ServletResponse 的子类，可通过该子类设置 HTTP 响应头或向客户端写入 Cookie 对象。

（1）ServletResponse 接口

Servlet 容器在接收到客户请求时，除创建 ServletRequest 对象用于封装客户的请求信息之外，还创建了一个 ServletResponse 对象用来封装响应信息。ServletResponse 接口的常用方法见表 6-5。

表 6-5　ServletResponse 接口常用方法

方法	说明
getWriter()	返回 PrintWrite 对象，用于向客户端发送文本
getCharacterEncoding()	返回在响应中发送的正文所使用的文字编码
setCharacterEncoding()	设置发送到客户端的响应的字符编码
setContentType(String type)	设置发送到客户端的响应的内容类型，此时响应的状态属于尚未提交

（2）HttpServletResponse 接口

HttpServletResponse 接口继承自 ServletResponse 接口，用于对客户端的请求进行响应。其除具有 ServletResponse 接口的常用方法之外，还增加了新的方法。HttpServletResponse 接口常用方法见表 6-6。

表 6-6　HttpServletResponse 接口常用方法

方法	说明
addHeader(String name,String value)	增加一个 Cookie 到响应中，这个方法可多次调用，设置多个 Cookie
addHeader(String name,String value)	将一个名称为 name，值为 value 的响应报头添加到响应中
sendRedirect(String location)	发送一个临时的重定向响应到客户端，以便客户端访问新的 URL
encodeURL(String url)	使用 sessionID 对用于重定向的 URL 进行编码

4.Servlet 配置方法

Servlet 有两种配置方法，一种是使用 web.xml 配置文件进行配置，另一种是使用注解 @WebServlet("/xxx") 的方法进行配置。

（1）使用 web.xml 配置

使用配置文件 web.xml 进行 Servlet 配置，该文件位于 WEB-INF 目录下，如图 6-4 所示。

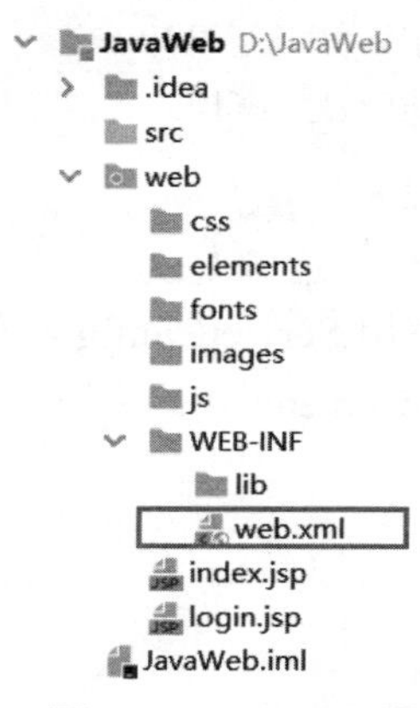

图 6-4　Servlet 配置

web.xml 文件的内容由 J2EE 规范描述，每个 Web 应用都需要有这个文件。该文件的主要用途就是向容器描述如何部署这个 Web 应用程序，标明 Servlet 的映射关系。

当浏览器传送请求到后端时，web.xml 文件的执行顺序如下。

第一步：从当前工程中的路径与 servlet-mapping 标签中的 url-pattern 的标签值进行匹配。

第二步：根据这个映射值，找到 servlet-mapping 标签中的 servlet-name 的值与 servlet 标签中的 servlet-name 进行匹配，找到 servlet 标签中的 servlet-class 标签中对应 servlet 类的 src 文件夹下的全路径。

第三步：调用并执行相应的 servlet 类。

例如：配置名称为 myServlet 的 Servlet，web.xml 文件应进行如下配置。

```
<servlet>
<servlet-name>myServlet</servlet-name>//servlet 自定义唯一名称
<servlet-class>demo.web.servlet.MyServlet</servlet-class>//servlet 类所在路径
</servlet>
<servlet-mapping>
    <servlet-name>myServlet</servlet-name>// 与 < servlet > 的 < servlet-name > 设置相同
<url-pattern>/myServlet</url-pattern>//servlet 的映射路径（访问 servlet 的名称）
</servlet-mapping>
```

可以使用 <init-param> 标签传递 Servlet 参数。一个 Servlet 可以配置一个或多个初始化参数。在应用程序中，可以使用 Servlet 的 getInitParameter(String param) 来读取初始化 param 对应的参数；若要读取所有的初始化参数名称，则可以使用 getInitParameterNames() 方法获得所有的参数名称，类型为枚举（Enumeration）。

例如：在 servlet 中定义 servlet 参数 charSetContent，代码如下。

```
<servlet>
  ……
<init-param>
<param-name>charSetContent</param-name>// 参数名称
<param-value>utf-8</param-value>// 参数值
</init-param>
</servlet>
```

上述的代码定义完成后，可以使用 ServletConfig 对象中的 getInitParameter() 方法读取初始化参数。例如：调用参数 charSetContent，代码如下。

```
public void init(ServletConfig config) throws ServletException
{
String initParam = config.getInitParameter("charSetContent");
    System.out.println(initParam);
}
```

（2）在 Servlet 类使用注解

Servlet 配置方法可以通过在对应的 Servlet 类中添加 Servlet 注解实现，从浏览器发送请求时是用当前“工程”下的路径，去对应 Servlet 类的上面寻找是否存在对应 URL 名称的 @WebServlet 注解，如若存在，调用并执行对应的 Servlet 类。

例如：Servlet 名称为 DemoJavaweb，在对应的 Servlet 类上使用 @WebServlet("/DemoJavaweb ") 注解即可。

```
@WebServlet("/DemoJavaweb ")
public class DemoJavaweb extends HttpServlet{
}
```

这两种配置方式的共同点是都能完成对 Servlet 的访问，区别如表 6-7 所示。

表 6-7　两种配置方法之间的区别

区别	优点	缺点
web.xml 配置	集中管理各 Servlet 类路径的映射路径，便于修改和管理	代码多，可读性不强，不易于理解
注解配置	代码少，可读性强，易于理解	如果大量使用 Servlet 注解，Servlet 类文件数量过多，不便于查找和修改

推荐使用 web.xml 文件配置，在方便管理的同时，更能加深对数据流动、系统流程的理解。

技能点三　Servlet 实例

【实例】实现 Servlet 获取表单信息。

Servlet 可以完成 HTML 表单数据的获取，只需在 Servlet 中调用 HttpServletRequest 中的 getParameter() 方法，并获取属性名即可。

例如在前端页面中填入用户名、密码等信息并单击“确定”之后，即将信息传送至 Servlet 中，效果如图 6-5（a）所示。

在 Servlet 中使用 PrintWriter 构造输出对象 Out，使用 Out 对象编辑 HTML 页面，并使用 getParameter() 方法获取名称为“username”和“password”表单对象的值，效果如图 6-5

(b)所示。

用户名: 密码: 确定

(a)

您所输入的信息为:

- 用户名: admin
- 密码: 123456

(b)

图 6-5 Servlet 读取表单实例

login 页面使用 form 标签向 LoginServlet 传递名为“username”和“password”的两个数值,具体如示例代码 6-1 所示。

示例代码 6-1:login.jsp

```
<%@ page language="java" contentType="text/html; charset=UTF-8"
    pageEncoding="UTF-8"%>
<html>
<head>
  <meta http-equiv="Content-Type" content="text/html; charset=UTF-8">
  <title>Login</title>
</head>
<body>
<form action="LoginServlet" method="post">
  用户名:<input name="username" type="text"/>
  密码:<input name="password" type="password"/>
  <input name="submit" type="submit" value=" 确定 " />
</form>
</body>
</html>
```

Servlet 中,由于 form 表单传送方法为 post,需要重载 doPost() 方法。使用 setCharacter-Encoding() 将页面编码设置为 OTF-8 以防止在传递中文时发生乱码,用 PrintWriter 构造输出对象 Out,使用 Out 对象编辑 HTML 页面,用 getParameter() 方法获取前端页面传递的名为“username”和“password”的数值，具体如示例代码 6-2 所示。

示例代码 6-2：LoginServlet.java

```
package com.loginservlet;
import java.io.IOException;
import java.io.PrintWriter;

import javax.servlet.ServletException;
import javax.servlet.annotation.WebServlet;
import javax.servlet.http.HttpServlet;
import javax.servlet.http.HttpServletRequest;
import javax.servlet.http.HttpServletResponse;

/**
 * Servlet implementation class LoginServlet
 */
public class LoginServlet extends HttpServlet {
    private static final long serialVersionUID = 1L;
    /**
     * @see HttpServlet#doGet(HttpServletRequest request, HttpServletResponse response)
     */
        protected void doGet(HttpServletRequest request, HttpServletResponse response)
throws ServletException, IOException {
        // TODO Auto-generated method stub
        doPost(request, response);
    }

    /**
        * @see HttpServlet#doPost(HttpServletRequest request, HttpServletResponse re-
sponse)
     */
        protected void doPost(HttpServletRequest request, HttpServletResponse response)
throws ServletException, IOException {
        // TODO Auto-generated method stub
        request.setCharacterEncoding("utf-8");// 设置页面编码
        response.setContentType("text/html;charset=utf-8");// 设置头部和编码
```

```
        response.setCharacterEncoding("utf-8");// 设置页面编码
        PrintWriter out = response.getWriter();// 构造输出对象 out
        out.println("<html>");// 使用 out 对象编写 HTML 页面
        out.println("<head><title>servlet</title></head>");
        out.println("<body>");
        out.println("<h3> 您所输入的信息为：</h3>");
        out.println("<li> 用户名: "+request.getParameter("username"));// 获取 username 属
性值
        out.println("<li> 密码: "+request.getParameter("password"));// 获取 password 属性
值
        out.println("</body></html>");
    }

}
```

编写 LoginServlet 页面之后，需要配置 web.xml 文件或者使用注解 @WebServlet("/LoginServlet")，在本实例中使用 web.xml 文件配置，具体如示例代码 6-3 所示。

示例代码 6-3：web.xml

```
<servlet>
  <servlet-name>LoginServlet</servlet-name>// 唯一名称 LoginServlet
  <servlet-class>com.loginservlet.LoginServlet</servlet-class>// 配置路径包名 + 类名
</servlet>
<servlet-mapping>
  <servlet-name>LoginServlet</servlet-name>// 与 < servlet-name > 名称一致
  <url-pattern>/Login</url-pattern>// 自定义访问名称
</servlet-mapping>
```

【实例】实现 Servlet 读取 Session 数据。

应用 Servlet 获取 Session 数据，在 Servlet 中使用 PrintWriter 构造输出对象 Out 并编辑 HTML 页面，然后使用 Request 方法获取请求信息，最终使用 Session 对象获取 Session 信息。

首次访问该页面，没有请求信息， Session 标识为首次创建，并获取 SessionID 创建时间等信息，效果如图 6-6（a）所示。

第二次访问该页面时，请求获取到 SessionID 号，使用 Cookie 在浏览器存储用户信息，并激活 Session。Session 信息中，首次创建被标识为 false，创建时间也有所不同，效果如图 6-6（b）所示。

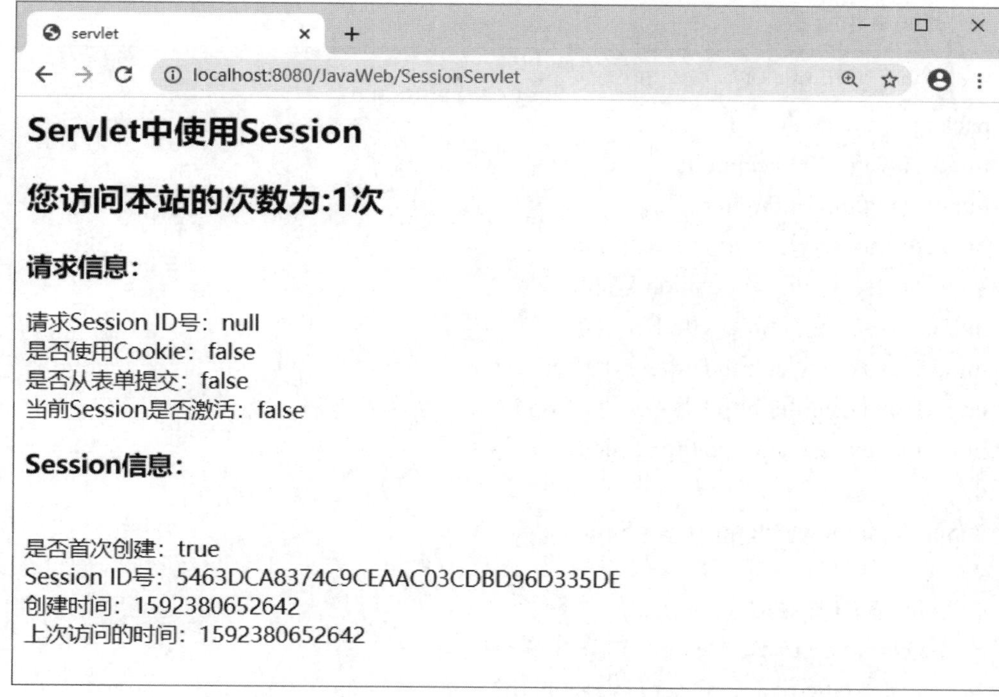

（a）

（b）

图 6-6　Servlet 读取 Session 信息实例

SessionServlet 重写了 doGet() 方法，使用 PrintWriter 拼接 HTML 代码显示在页面上，具

体如示例代码 6-4 所示。

示例代码 6-4: SessionServlet.java

```
package com.sessionservlet;
import java.io.IOException;
import java.io.PrintWriter;
import javax.servlet.ServletException;
import javax.servlet.annotation.WebServlet;
import javax.servlet.http.HttpServlet;
import javax.servlet.http.HttpServletRequest;
import javax.servlet.http.HttpServletResponse;
import javax.servlet.http.HttpSession;
/**
 * Servlet implementation class SessionServlet
 */
@WebServlet("/SessionServlet")
public class SessionServlet extends HttpServlet {
    private static final long serialVersionUID = 1L;
    /**
     * @see HttpServlet#doGet(HttpServletRequest request, HttpServletResponse re-
sponse)
 */
    protected void doGet(HttpServletRequest request, HttpServletResponse response)
throws ServletException, IOException {
    // TODO Auto-generated method stub
    HttpSession session = request.getSession();// 获取 session 对象
    request.setCharacterEncoding("utf-8");// 设置页面中文的编码
    response.setContentType("text/html;charset=utf-8");
            response.setCharacterEncoding("utf-8");
    PrintWriter out = response.getWriter();
    out.println("<html>");
    out.println("<head><title>servlet</title></head>");
    out.println("<body>");
    out.println("<p>");
    out.println("<h2>Servlet 中使用 Session<h2>");
    Integer count = (Integer)session.getAttribute("counter");// 获取 session 中的 counter
数值
    if (count==null) {// 若 count 为 null 则表示第一次访问该页面
    count = new Integer(1); // 设置 count 变量赋值为 1，记录访问页面的次数
```

```
        }
        else
        count = new Integer(count.intValue()+1); // 若 count 不为 null 则在原有基础上加 1
        session.setAttribute("counter", count);// 将计数情况重新赋值给 counter
        out.println(" 您访问本站的次数为 :"+count+" 次 </p>");
        out.println("<h3> 请求信息：</h3>");
        out.println(" 请求 Session ID 号："+request.getRequestedSessionId());
        out.println("<br> 是否使用 Cookie："+request.isRequestedSessionIdFromCookie());
        out.println("<br> 是否从表单提交："+request.isRequestedSessionIdFromURL());
                out.println("<br> 当前 Session 是否激活："+request.isRequestedSessionId-
Valid());
        out.println("<h3>Session 信息：</h3>");
        out.println("<br> 是否首次创建："+session.isNew());
        out.println("<br>Session ID 号："+session.getId());
        out.println("<br> 创建时间："+session.getCreationTime());
        out.println("<br> 上次访问的时间："+session.getLastAccessedTime());
        out.println("</body></html>");
        }
        /**
         * @see HttpServlet#doPost(HttpServletRequest request, HttpServletResponse re-
sponse)
    */
        protected void doPost(HttpServletRequest request, HttpServletResponse response)
throws ServletException, IOException {
        // TODO Auto-generated method stub
        doGet(request, response);
        }
    }
```

在本实例中，使用了注解的方式配置 Servlet，代码如下。

```
@WebServlet("/SessionServlet")// 使用注解的方法配置 Servlet
public class SessionServlet extends HttpServlet {
    ……
}
```

本实例使用 web.xml 方式配置 Servlet，代码如下。

```
<servlet>
  <servlet-name> SessionServlet </servlet-name>
  <servlet-class>com. sessionservlet. SessionServlet </servlet-class>
```

```
</servlet>
<servlet-mapping>
  <servlet-name> SessionServlet </servlet-name>
  <url-pattern>/SessionServlet </url-pattern>
</servlet-mapping>
```

技能点四 Servlet 过滤器

1.Servlet 过滤器

Servlet 过滤器是在 Java Servlet 规范 2.3 中定义的，它能够对 Servlet 容器请求和响应进行检查和修改。Servlet 过滤器本身不会产生请求和响应，只会在访问目标页面或是 Servlet 资源时，对 Request 对象和 Response 对象进行检查，其流程如图 6-7 所示。

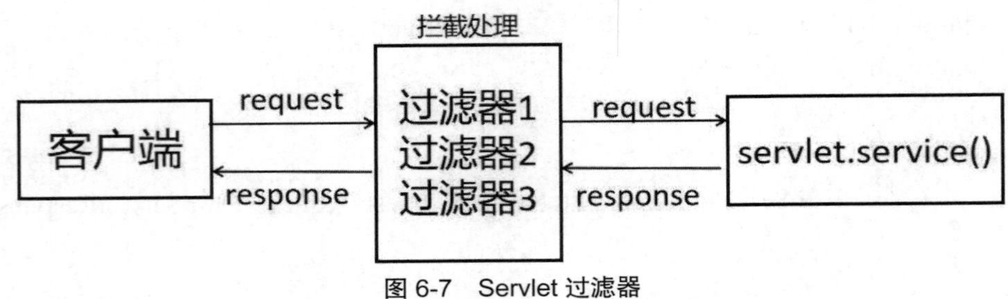

图 6-7 Servlet 过滤器

Servlet 过滤器的特点如下。

1）Servlet 过滤器可以检查和修改 ServletRequset 和 ServletRequest 对象。

2）Servlet 过滤器可以被指定和特定的 URL 关联，只有当用户请求特定的目标 URL 时，才会触发。

3）Servlet 过滤器可以被串联在一起，同时修改请求和响应。

Servlet 过滤器的工作过程，包括如下几个步骤。

第一步：用户访问 Web 资源时，发送请求会首先经过过滤器。

第二步：由过滤器对请求的数据进行处理。

第三步：经过首次处理之后的请求数据被发送至目标资源进行处理。

第四步：目标资源经过处理响应后再次被发送至响应过滤器进行处理。

第五步：经过过滤器的处理之后，将响应返回给客户端进行显示。

2.Servlet 过滤器接口常用方法

过滤器的工作流程如下。

第一步：实现 javax.servlet.Filter 接口。

第二步：实现 init() 方法，完成过滤器的初始化。

第三步：实现 doFilter() 方法，完成对于请求或者响应的过滤处理。

第四步：调用 FilterChain 接口对象的 doFilter() 方法，并向后续的过滤器传递请求或响应。

第五步：实现 destroy() 方法，销毁过滤器。

在以上工作流程中运用到了 Servlet 过滤器的 init()、doFilter()、destroy() 方法，这些方法的详情如表 6-8 所示。

表 6-8 Servlet 过滤器接口常用方法

方法	说明
init(FilterConfig)	初始化方法，服务器启动时会创建过滤器的实例化对象，并调用 init() 方法，完成初始化方法，过滤器对象只会创建一次，init() 方法也只会执行一次。在此方法中可以读取 web.xml 文件中 Servlet 过滤器初始化参数
doFilter(ServletRequest,ServletResponse,FilterChain)	完成具体过滤操作的方法，在访问过滤器关联的 URL 时，Serlvet 容器将先调用过滤器的 doFilter() 方法，FilterChain 参数用于访问后续的过滤器
destroy()	Servlet 过滤器销毁前调用该方法，可释放 Servlet 过滤器占用的资源

3.Servlet 过滤器实例

网上书城项目使用 Servlet 过滤器验证用户登录信息。

当通过修改地址栏 URL 访问首页跳过登录步骤时，会弹出提示框，跳转至登录界面，效果如图 6-8 所示。

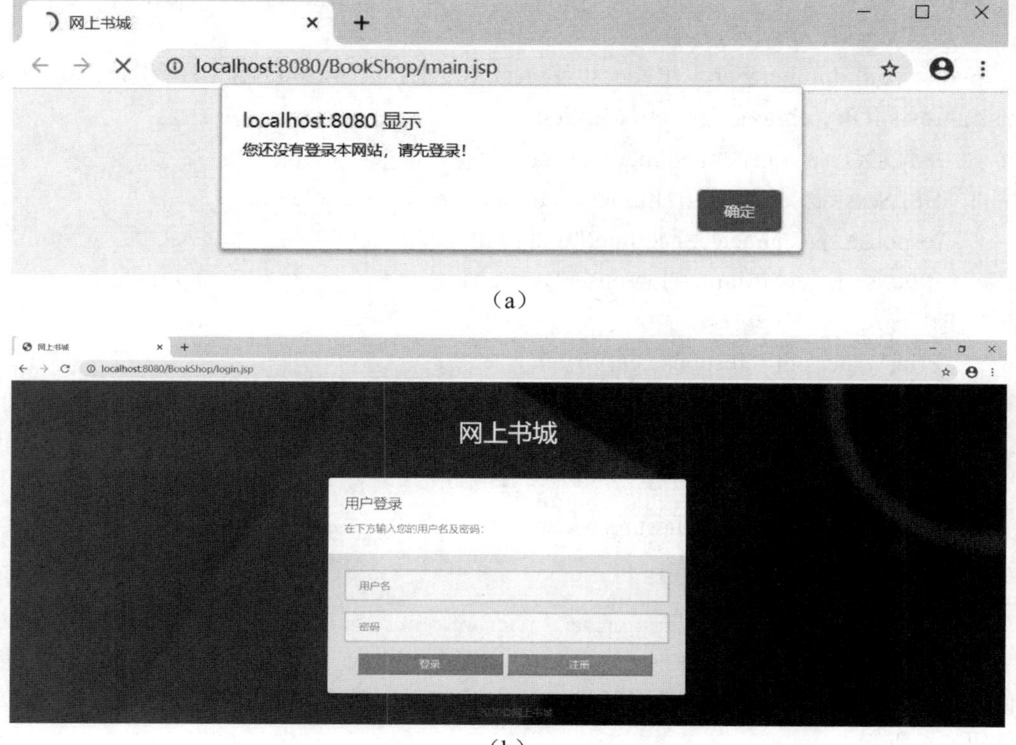

（a）

（b）

图 6-8 Servlet 过滤器实例

具体步骤如下。

第一步：在 util 文件夹下创建 LoginFilter 文件。其中，过滤器的逻辑为使用 session.getAttribute("loginuser") 获取已经登录用户的 Session 对应的属性值，并判断是否为 null，若为 null 则使用 PrintWriter 方法将 JavaScript 代码写入页面，设置异常弹出框和页面跳转，如示例代码 6-5 所示。

示例代码 6-5：LoginFilter.java

```
package com.xt.util;
import com.sun.deploy.net.HttpResponse;
import java.io.IOException;
import java.io.PrintWriter;
import javax.servlet.Filter;
import javax.servlet.FilterChain;
import javax.servlet.ServletException;
import javax.servlet.ServletRequest;
import javax.servlet.ServletResponse;
import javax.servlet.http.HttpServlet;
import javax.servlet.http.HttpServletRequest;
import javax.servlet.http.HttpServletResponse;
import javax.servlet.http.HttpSession;
    @Override
public void doFilter(ServletRequest request, ServletResponse response, FilterChain filterChain)throws IOException, ServletException {
    // TODO Auto-generated method stub
    HttpSession session = ((HttpServletRequest)request).getSession();
    response.setCharacterEncoding("UTF-8");
    if (session.getAttribute("loginuser")==null)
    {
    PrintWriter out = response.getWriter();
    out.println("<script  language=javascript>alert('您还没有登录本网站，请先登录！');window.location.href='/BookShop/login.jsp';</script>");
    }
    filterChain.doFilter(request,response);
    }
}
```

第二步：设置 web.xml 文件，<url-pattern> 标签用来指定需要进行过滤的 JSP 文件，也可以使用“/*”或“/ 文件夹名称 /*”的形式来指定过滤根文件夹或指定文件夹中的文件，如示例代码 6-6 所示。

示例代码 6-6：web.xml

```
<filter>
    <filter-name>LoginFilter</filter-name>// 名称唯一
    <filter-class>com.xt.util.LoginFilter</filter-class>// 资源地址包名 + 类名
</filter>
<filter-mapping>
   <filter-name>LoginFilter</filter-name>// 与上方 <filter-name> 标签一致
    <url-pattern>/main.jsp</url-pattern>// 设置首页过滤
<url-pattern>/shopping.jsp</url-pattern>// 设置购物车页面过滤
<url-pattern>/orderlist.jsp</url-pattern>// 设置订单页面过滤
</filter-mapping>
```

注意：不能将登录页面包含到登录过滤器中，否则就会出现用户无法登录的现象。

技能点五　Servlet 监听器

1.Servlet 监听器概述

Servlet 监听器是在 Servlet2.3 规范中与 Servlet 过滤器一起引入的，主要是对于 Web 项目进行监听和控制，监听某个对象的状态变化的组件，监听的范围在 request、Session 和 ServletContext 中。

2.Servlet 监听器常用接口以及方法

在 Servlet 监听器中，通常会监听 Servlet 上下文对象或者 Session 对象，它常用的接口方法见表 6-9。

表 6-9　Servlet 监听器常用接口方法

接口名称	说明
javax.servlet.ServletContextListener	在 Servlet 上下文对象初始化或者销毁时，得到通知
javax.servlet.ServletContextAttributeListener	在 Servlet 上下文中的属性列表发生变化时，得到通知
javax.servlet.http.HttpSessionListener	在 Session 创建后或者失效前，得到通知
javax.servlet.http.HttpSessionActivationListener	绑定到 Session，当 Session 发生变化时，通知该对象
javax.servlet.http.HttpSessionAttributeListener	在 Session 中的属性发生变化时，得到通知
javax.servlet.ServletRequestListener	请求对象初始化或者销毁时，得到通知
javax.servlet.ServletRequestAttributeListener	请求对象中属性发生变化时，得到通知
javax.servlet.http.HttpSessionBindingListener	使一个对象在绑定或者解绑 Session 时，得到通知

3.Servlet 上下文监听

Servlet 上下文监听可以监听 ServletContext 对象的创建、删除以及对属性的添加、删除和修改操作。在 JSP 文件中，Application 中 ServletContext 的实例，由 JSP 容器默认创建。使用 getServletContext() 方法得到 ServletContext 实例。

初始化阶段，在 web.xml 中经由 <context-param> 标签所设定的变量，在容器启动时，可在 Application 范围内使用 getInitParameter() 方法获取相对应的属性值。

例如，在 web.xml 中进行如下配置。

```
<context-param>
<param-name>user</param-name>
<param-value>john</param-value>
</context-param>
```

在 JSP 页面中获取名为“user”的变量。

```
String name = (String)application.getInitParameter("user")
```

在 Servlet 中获取名为“user”的变量。

```
String name = (String)ServletContext.getInitParameter("user")
```

Servlet 上下文监听需要用到以下两个接口。

1）ServletContextListener 接口，该接口存放在 javax.servlet 包内，主要实现监听 ServletContext 的创建和删除，见表 6-10。

表 6-10 ServletContextListener 常用方法

方法	说明
contextInitialized(ServletContextEvent event)	通知正在收听的对象，应用程序已经被加载以及初始化
contextDestroyed(ServletContextEvent event)	通知正在收听的对象，应用程序已经被关闭

2）ServletAttributeListener 接口，该接口存放在 javax.servlet 包内，主要实现监听 ServletContext 属性的添加、删除和修改，具体内容见表 6-11。

表 6-11 ServletAttributeListener 常用方法

方法	说明
attributeAdded(ServletContextAttributeEvent event)	当有对象加入 Application 的范围时，通知正在收听的对象
attributeReplaced(ServletContextAttributeEvent event)	当在 Application 的范围有对象取代另一个对象时，通知正在收听的对象
attributeRemoved(ServletContextAttributeEvent event)	当有对象从 Application 的范围移除时，通知正在收听的对象

4. HTTP 会话监听

HTTP 会话监听信息会用到四个接口，具体内容如下。

（1）HttpSessionBindingListener 接口

当类实现了 HttpSessionBindingListener 接口后，只要对象加入 Session 范围（即调用

HttpSession 对象的 setAttribute 方法）或从 Session 范围中移出（即调用 HttpSession 对象的 removeAttribute 方法或 Session Time out）时，容器会分别自动调用表 6-12 所示方法。

表 6-12　HttpSessionBindingListener 常用方法

方法	说明
valueBound(HttpSessionBindingEvent event)	当有对象加入 Session 的范围时，会被自动调用
valueUnbound(HttpSessionBindingEvent event)	当有对象从 Session 的范围内移除时，会被自动调用

（2）HttpSessionAttributeListener 接口

该接口监听 HttpSession 中的属性操作，提供表 6-13 所示三种方法。

表 6-13　HttpSessionAttributeListener 常用方法

方法	说明
attributeAdded(HttpSessionBindingEvent se)	Session 增加一个属性时激发
attributeRemoved(HttpSessionBindingEvent se)	Session 删除一个属性时激发
attributeReplaced(HttpSessionBindingEvent se)	Session 属性被重新设置时激发

（3）HttpSessionListener 接口

该接口监听 HttpSession 的操作，提供表 6-14 所示两种方法。

表 6-14　HttpSessionListener 常用方法

方法	说明
sessionCreated(HttpSessionEvent se)	当创建一个 Session 时激发
sessionDestroyed(HttpSessionEvent se)	当销毁一个 Session 时激发

（4）HttpSessionActivationListener 接口

该接口监听 HTTP 会话 activate、passivate 状态，提供表 6-15 所示两种方法。

表 6-15　HttpSessionActivationListener 常用方法

方法	说明
sessionDidActivate(Http Session Event se)	通知正在收听的对象，Session 状态变为有效状态
sessionWillPassivate(Http Session Event se)	通知正在收听的对象，Session 状态变为无效状态

5.Servlet 请求监听

Servlet 请求监听是在 Servlet2.4 规范中新增的技术，可以监听客户端的请求。一旦能够在监听程序中获取客户端的请求，就可以对请求进行统一处理。

（1）ServletResquestListener 接口

该接口监听 HttpResquset 中的属性操作，并提供表 6-16 所示两种方法。

表 6-16 ServletResqestListener 常用方法

方法	说明
requestInitalized(ServletResqestEvent event)	通知正在收听的对象，ServletResquset 已经被加载以及初始化
requestDestroyed(ServletResqestEvent event)	通知正在收听的对象，ServletResquset 已经被关闭

（2）ServletResqestAttributeListener 接口

该接口监听 HttpSession 中的属性操作，并提供表 6-17 所示三种方法。

表 6-17 ServletResqestAttributeListener 常用方法

方法	说明
attributeAdded(ServletResqestAttributeEvent event)	当有对象加入 resquest 的范围时，通知正在收听的对象
attributeReplaced(ServletResqestAttributeEvent event)	当在 resquest 的范围有对象取代另一个对象时，通知正在收听的对象
attributeRemoved(ServletResqestAttributeEvent event)	当有对象从 resquest 的范围移除时，通知正在收听的对象

6.Servlet 监听器实例

网上书城项目应用 HttpSessionListener 接口实现对在线人数的统计。在导航栏上标有“当前在线人数”字样的在线人数统计功能，效果如图 6-9 所示。

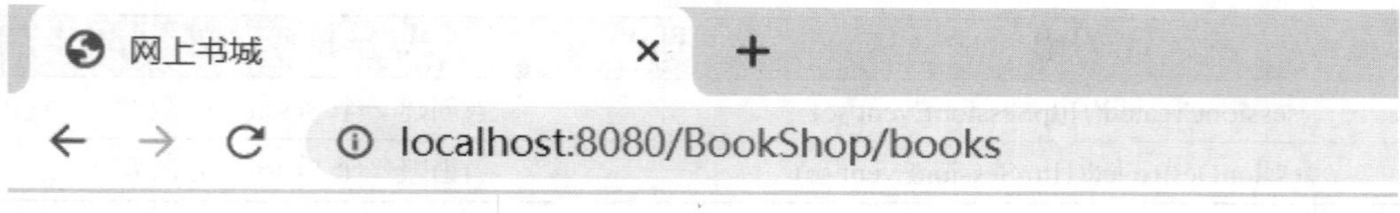

图 6-9 监听器实例

第一步：在 util 文件夹下创建 OnlineCounter 文件，编写计数功能的 Java 代码，具体如示例代码 6-7 所示。

示例代码 6-7：OnlineCounter.java

```
package com.xt.util;
```

```
public class OnlineCounter {
    private static long online = 0 ;// 定义静态变量 online
    public static long getOnline(){// 获取当前在线人数
        return online;
    }
    public static void add(){// 计数功能
        online++;
    }
   public static void reduce(){// 计数功能
        online--;
    }
}
```

第二步：在 util 文件夹下创建 OnlineCounterListener 文件，在其中应用 HttpSessionListener 接口，在 OnlineCounterListener 类中重写 sessionCreated() 方法，在其中调用 OnlineCounter 中的 add() 和 reduce() 方法。当新的 Session 被创建时，调用 add() 方法，增加统计的人数，当 Session 被销毁时，调用 reduce() 方法，在线人数减 1，具体如示例代码 6-8 所示。

示例代码 6-8：OnlineCounterListener.java

```
package com.xt.util;

import javax.servlet.http.HttpSessionEvent;
import javax.servlet.http.HttpSessionListener;

public class OnlineCounterListener implements HttpSessionListener {
    public void sessionCreated(HttpSessionEvent httpSessionEvent)
    {
        OnlineCounter.add();// 创建 Session 时，调用 OnlineCounter 计数功能中的 add()
方法在线人数加 1
    }
    public void sessionDestroyed(HttpSessionEvent httpSessionEvent)
    {
        OnlineCounter.reduce();// 销毁 Session 时，调用 OnlineCounter 计数功能中的 re-
duce() 方法在线人数减 1

    }
}
```

第三步：配置 web.xml 文件，在 <listener-class> 标签中新增 <listener> 标签，内容为监听

器具体位置（包名＋类名），具体如示例代码 6-9 所示。

示例代码 6-9：web.xml

```
<listener>
  <listener-class>com.xt.util.OnlineCounterListener</listener-class>
</listener>
```

1. 网上书城登录模块后台逻辑编写

在第二章中，编写完成了登录模块的前端页面。在本章中要实现前端页面与后台的数据交互，具体步骤如下。

第一步：在 src 文件夹下的 com.xt 包中创建 service 文件夹和 servlet 文件夹，依次存放 service 文件和 servlet 文件。其中，Servlet 过滤器和监听器属于工具文件，放置在 util 文件夹下，如图 6-10 所示。

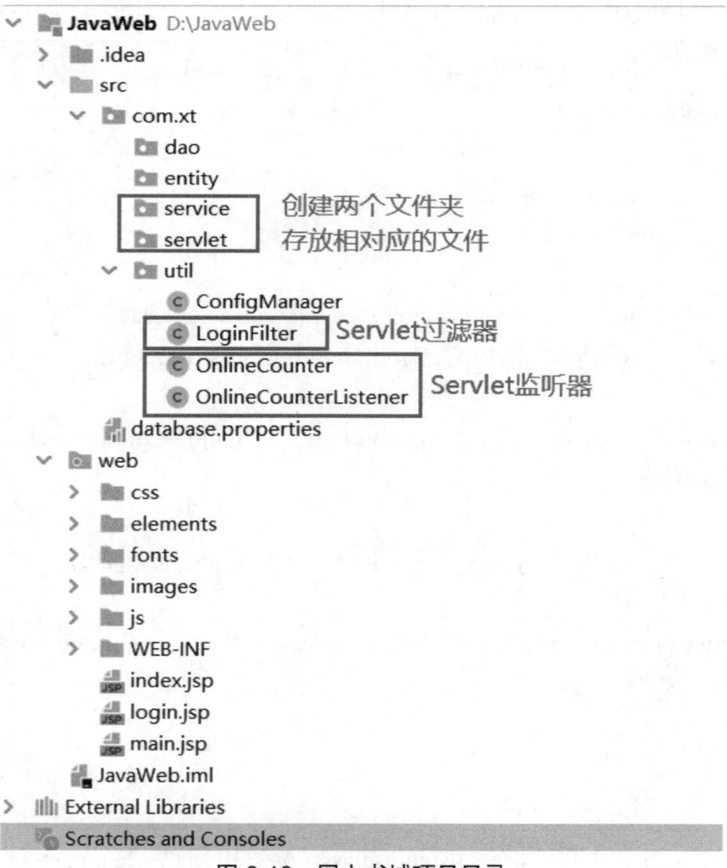

图 6-10 网上书城项目目录

第二步：编写 LoginServlet 文件。重写 doPost() 方法通过 userbiz.checkLogin(username, password) 方法查询用户是否存在，对于存在的用户将用户信息存入 Session 并跳转至主页，对于不存在的用户则跳转至登录页面，重新登录，具体如示例代码 6-10 所示。

示例代码 6-10：LoginServlet.java

```
package com.xt.servlet;
import java.io.IOException;
import java.io.PrintWriter;
import javax.servlet.ServletException;
import javax.servlet.http.HttpServlet;
import javax.servlet.http.HttpServletRequest;
import javax.servlet.http.HttpServletResponse;
import com.xt.service.UserService;
/*
 * 登录
 */
public class  LoginServlet extends HttpServlet {
    private UserService userbiz = null;
    @Override
    public void init() throws ServletException {
    userbiz = new UserService();
    }
    @Override
    protected void doPost(HttpServletRequest req, HttpServletResponse resp)
    throws ServletException, IOException {
    String username = req.getParameter("username");
    String password = req.getParameter("password");
    // 登录操作
// 通过 checkLogin 查询用户是否存在
    int uid = userbiz.checkLogin(username, password);
    // 登录成功
    if(uid>0){
// 查询到用户则将用户信息存入 Session
    req.getSession().setAttribute("uid", uid);
    req.getSession().setAttribute("loginuser", username);
    resp.sendRedirect("books");
    }
    // 登录失败
    else{
```

```
        // 用户不存在则使用 println 拼接 HTML 语句重新登录
        resp.setContentType("text/html; charset=utf-8");
        PrintWriter pw = resp.getWriter();
        pw.println("<script type=\"text/javascript\">");
        pw.println("alert(\" 登录失败！ 请重新登录！！ \");");
        pw.println("open(\"login.jsp\",\"_self\");");
        pw.println("</script>");
        pw.close();
        }
    }
```

第三步：编写 UserDao 类，首先编写查询用户信息的 SQL 语句，再调用 doSelect(sql,params) 方法，查询是否存在该登录用户，具体如示例代码 6-11 所示。

示例代码 6-11：UserDao.java

```
package com.xt.dao;
import java.util.ArrayList;
import java.util.HashMap;
import java.util.List;
import com.xt.entity.UserInfo;
public class UserDao extends BaseDao {
    public List<HashMap> query(UserInfo userinfo) {
    String sql = "SELECT * FROM userinfo WHERE username =? AND password = ?";
    Object[] params = {userinfo.getUsername(),userinfo.getPassword()};
    return doSelect(sql,params);
    }
}
```

第四步：编写 UserService 类，调用 checkLogin(String username, String password) 方法查询用户是否存在，具体如示例代码 6-12 所示。

示例代码 6-12：UserService.java

```
package com.xt.service;
import java.util.HashMap;
import java.util.List;
import com.xt.dao.UserDao;
import com.xt.entity.UserInfo;
public class UserService {
    private UserDao userdao = new UserDao();
    /*
```

```
* 执行登录请求操作
    */
        public int checkLogin(String username, String password) {
        UserInfo userInfo = new UserInfo();
        userInfo.setUsername(username);
        userInfo.setPassword(password);
        List<HashMap> list = userdao.query(userInfo);
        return list.size()>0?(int)list.get(0).get("uid"):-1;
        }
    }
```

第五步：编写公共方法 BaseDao 类，打开数据库链接，通过传递的 SQL 参数以及从前端获取的数据，查询对应数据，返回一个 list 集合，并关闭数据库链接，具体如示例代码 6-13 所示。

示例代码 6-13：BaseDao.java

```
    package com.xt.dao;
    import java.sql.*;
    import java.util.ArrayList;
    import java.util.HashMap;
    import java.util.List;
    import java.util.Map;
    import com.xt.util.ConfigManager;
    public class BaseDao {
        protected Connection conn = null;
        protected PreparedStatement ps = null;
        protected ResultSet rs = null;
        /*
    * 获取数据库链接
*/
        protected void openconnection(){// 与数据库链接的方法
        String driver=ConfigManager.getInstance().getString("jdbc.driver_class");
        String url=ConfigManager.getInstance().getString("jdbc.connection.url");
        String username=ConfigManager.getInstance().getString("jdbc.connection.username");
        String password=ConfigManager.getInstance().getString("jdbc.connection.password");
        try {
        Class.forName(driver);
        conn = DriverManager.getConnection(url, username, password);// 链接的核心方法
        } catch (ClassNotFoundException e) {
```

```
e.printStackTrace();
} catch (SQLException e) {
e.printStackTrace();
}
}
/*
* 执行查询的方法
*/
public List doSelect(String sql, Object[] params) {
List<Map> list = new ArrayList<Map>();
try {
openconnection();
ps = conn.prepareStatement(sql);
for (int i = 0; i < params.length; i++) {
ps.setObject(i + 1, params[i]);
}
rs = ps.executeQuery();
ResultSetMetaData metaData = rs.getMetaData();
int cols_len = metaData.getColumnCount();
while (rs.next()) {
Map map = new HashMap();
for (int i = 0; i < cols_len; i++) {
String cols_name = metaData.getColumnName(i + 1);
Object cols_value = rs.getObject(cols_name);
if (cols_value == null) {
cols_value = "";
}
map.put(cols_name, cols_value);
}
list.add(map);
}
} catch (Exception e) {
e.printStackTrace();
} finally {
closeResource();
}
return list;
}
```

```
        /*
* 关闭资源
*/
        protected boolean closeResource(){
        try {
        if(rs != null)
        rs.close();
        if(ps != null)
        ps.close();
        if(conn != null)
        conn.close();
        } catch (SQLException e) {
        e.printStackTrace();
        return false;
        }
        return true;
        }
    }
```

第六步：配置 web.xml 文件，具体如示例代码 6-14 所示。

示例代码 6-14：web.xml

```
<servlet>
  <description> 登录组件 </description>
  <servlet-name>login</servlet-name>
  <servlet-class>com.xt.servlet.LoginServlet</servlet-class>
</servlet>
<servlet-mapping>
  <servlet-name>login</servlet-name>
  <url-pattern>/login</url-pattern>
</servlet-mapping>
```

完成上述操作，网上书城项目登录模块后台逻辑部分就已经完成，结合第二章所设计的前端页面，效果如图 6-11 所示。

2. 网上书城注册模块编写

通过 Servlet 知识、JSP 基本知识完成网上书城注册功能的编写，效果如图 6-12 所示。

注册模块具有验证输入信息功能，对于不符合要求的填入信息进行驳回，如图 6-13 所示。

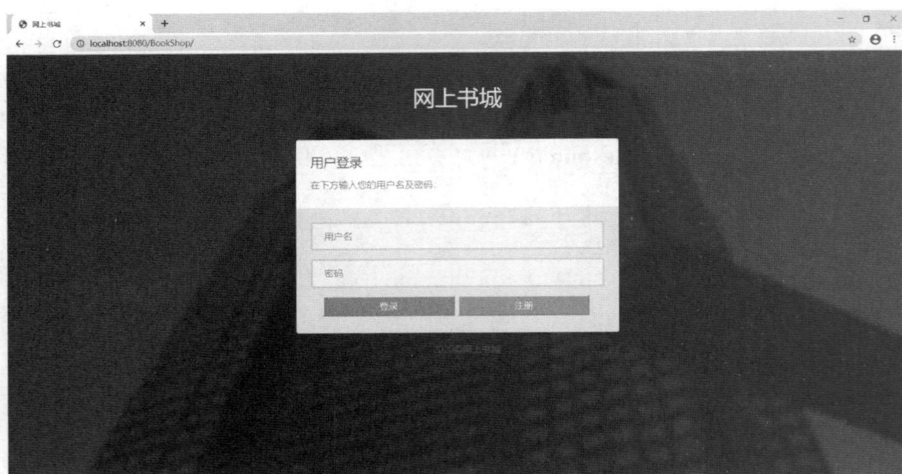

图 6-11　网上书城项目登录页面

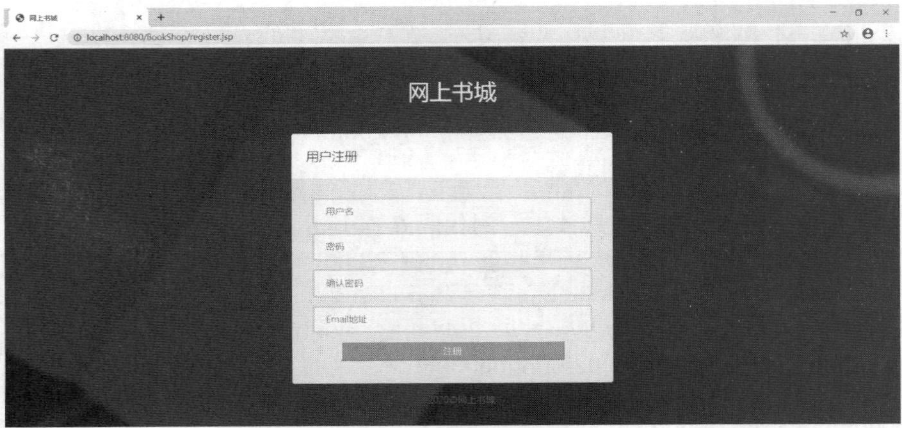

图 6-12　注册功能页面

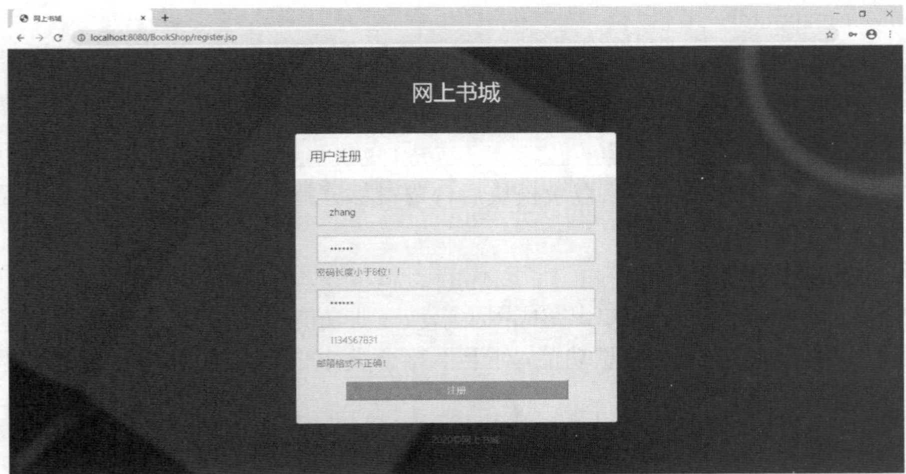

图 6-13　注册页面验证效果

修改为正确的信息之后跳转至注册成功页面，可通过单击“点击进入用户中心”跳转至主页面，如图 6-14 所示。

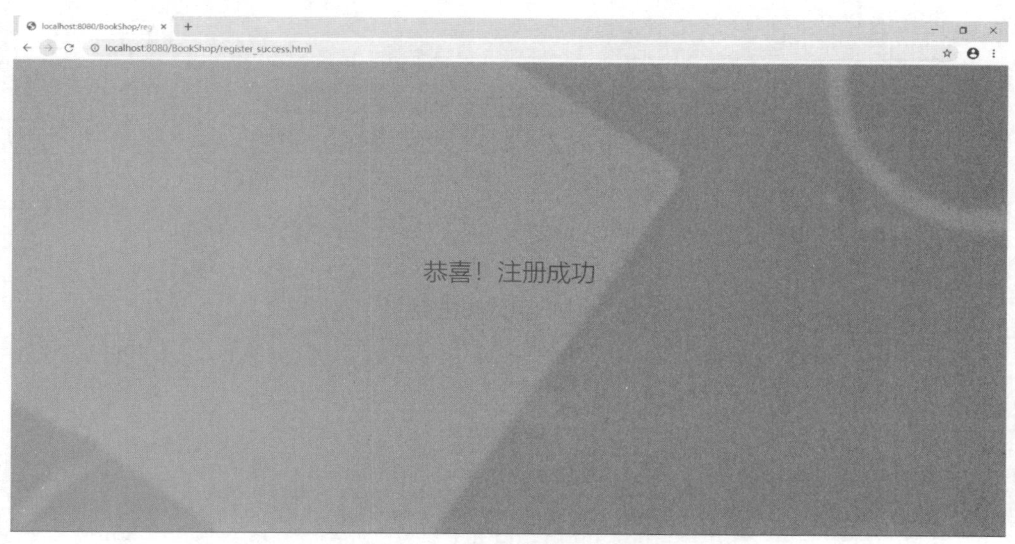

图 6-14　注册成功跳转效果

要实现上述效果，需要完成以下步骤。

第一步：编写前端 register.jsp 页面。运用 JavaScript 对表单数据进行验证，验证通过后传递至 RegistServlet 进行处理，具体如示例代码 6-15 所示。

示例代码 6-15：register.jsp

```
<%@ page language="java" import="java.util.*" contentType="text/html; charset=utf-8"%>
<jsp:include page="elements/index_head.jsp"></jsp:include>
<link rel="stylesheet" href="css/bootstrap1.min.css">
<link rel="stylesheet" href="css/font-awesome.min.css">
<link rel="stylesheet" href="css/form-elements.css">
<link rel="stylesheet" href="css/style2.css">
<script type="text/javascript">
    // 全部输入项非空检查
    // 检查用户名是否存在
    function isUsernameLegal(){
    var username = document.getElementById("username").value;
    var usernull = document.getElementById("usernull");
    if(username == null || username == ""){
    usernull.innerHTML = "<font color=\"red\"> 用户名不能为空！ </font>"
    return false;
    }
```

```
return true;
}
// 检查密码内容位数
function isPasswordLegal(){
var password = document.getElementById("password").value;
var pwdnull = document.getElementById("nullpassword");
if(password == null || password == ""){
pwdnull.innerHTML = "<font color=\"red\"> 密码不能为空！ </font>"
return false;
}else if(password.length < 8){
pwdnull.innerHTML = "<font color=\"red\"> 密码长度小于 8 位!! </font>"
return false;
}else
pwdnull.innerHTML = ""
return true;
}
// 检查密码是否一致
function isRepasswordLegal(){
var repassword = document.getElementById("rePassword").value;
var password = document.getElementById("password").value;
var pwdnull = document.getElementById("nullrePassword");
if(repassword == null || repassword == ""){
pwdnull.innerHTML = "<font color=\"red\"> 确认密码不能为空！ </font>"
return false;
}else if(repassword != password){
pwdnull.innerHTML = "<font color=\"red\"> 两次密码输入不一致！ </font>"
return false;
}else
pwdnull.innerHTML = ""
return true;
}
// 检查邮箱格式是否正确
function isEmailLegal(){
var email = document.getElementById("email").value;
var emailnull = document.getElementById("nullemail");
var regEx = /^[a-zA-Z0-9_-]+@[a-zA-Z0-9_-]+(\.[a-zA-Z_-]+)+$/;
if(email == null || email == ""){
emailnull.innerHTML = "<font color=\"red\"> 邮箱不能为空！ </font>"
```

```
    return false;
    }else if(!regEx.test(email)){
    emailnull.innerHTML = "<font color=\"red\"> 邮箱格式不正确！ </font>"
            return false;
            }else
 emailnull.innerHTML = ""
    return true;
    }
    function check(){
    var username = document.getElementById("username");
    if(!isUsernameLegal()){
    username.focus();
    return false;
    }else if(!isPasswordLegal()){
    username.focus();
    return false;
    }
    else if(!isRepasswordLegal()){
    username.focus();
    return false;
    }
    else if(!isEmailLegal()){
    username.focus();
    return false;
    }else{
    return true;
    }
    }
    function isUserExsit(){
    var info=document.getElementById("isExsit").value;
    if(info==null||info==""){
    return true;
    }else{
    usernull.innerHTML = "<font color=\"red\"> 当前用户名已被注册！ </font>";
    return false;
    }
    }
</script>
```

```
<body onload="isUserExsit()">
<div class="top-content">
    <div class="inner-bg">
    <div class="container">
    <div class="row book">
<div class="col-sm-8 col-sm-offset-2 text">
    <h1><strong> 网上书城 </strong></h1>
    </div>
    </div>
    <div class="row" id="register">
    <div class="col-sm-6 col-sm-offset-3 form-box">
    <div class="form-top">
    <div class="form-top-left">
    <h3> 用户注册 </h3>
    </div>
    </div>
    <div class="form-bottom">
    <%// 获取对应的 info 信息
    String info="";
    Object o=request.getAttribute("info");
    if(o!=null){
    info=o.toString();
    }
    %>
    <input type="hidden" id="isExsit" value="<%=info %>"/>
    <form method="post" action="regist" onsubmit="return check()" class="login-form">
//form 表单中的 action 属性指向 RegistServlet
    <div class="form-group">
    <label class="sr-only" for="username"> 用户名：</label>
    <input type="text" name="username" placeholder=" 用户名 " class="form-username form-control" id="username" onblur="isUsernameNull()" ><span id="usernull"></span><span id="alreadyExsits"></span>
    </div>
    <div class="form-group">
    <label class="sr-only" for="password"> 密      码：</label>
    <input type="password" name="password" placeholder=" 密码 " class="form-password form-control" id="password" onblur="isPasswordNull()"><span id="nullpassword"></span><span id="simplepassword"></span>
```

```
        </div>
        <div class="form-group">
        <label class="sr-only" for="rePassword"> 确认密码:</label>
        <input type="password" name="rePassword" placeholder=" 确认密码 " class="-form-password form-control" id="rePassword" onblur="isRepasswordLegal()"><span id="null-rePassword"></span><span id="uneq"></span>
        </div>
                <div class="form-group">
        <label class="sr-only" for="password">Email 地址:</label>
        <input class="form-password form-control" type="text" id="email" placeholder="E-mail 地址 " name="email" onblur="isEmailLegal()" /><span id="nullemail"></span><span id="errorInput"></span>
        </div>
        <button class="input-reg" type="submit" name="register"  style=" margin-left: 10%; width:80%;background:#19c880; color: white" > 注册 </button>
        </form>
        </div>
        </div>
        </div>
        </div>
        </div>
    </div>
    <script src="js/jquery-1.11.1.js"></script>
    <script src="js/bootstrap.min.js"></script>
    <script src="js/jquery.backstretch.js"></script>
    <script src="js/scripts.js"></script>
    </body>
    <jsp:include page="elements/main_bottom.html"/>
```

第二步:编写 register_success.html 页面,完成注册成功之后的跳转页面,具体如示例代码 6-16 所示。

示例代码 6-16:register_success.html

```
<!DOCTYPE html PUBLIC "-//W3C//DTD XHTML 1.0 Transitional//EN" "http://www.w3.org/TR/xhtml1/DTD/xhtml1-transitional.dtd">
<html xmlns="http://www.w3.org/1999/xhtml">
<head>
<meta http-equiv="Content-Type" contentType="text/html; charset=utf-8" />
<title></title>
```

```
    <link rel="stylesheet" href="css/bootstrap1.min.css">
    <link rel="stylesheet" href="css/font-awesome.min.css">
    <link rel="stylesheet" href="css/form-elements.css">
    <link rel="stylesheet" href="css/style2.css">
</head>
<body>
<div id="register">

    <div class="success" style="margin-top: 20%">

    <div class="information">

    <h1> 恭喜！注册成功 </h1>

    <h2><a href="login.jsp"> 点击进入用户中心 </a></h2>

    </div>
    </div>
</div>
<script src="js/jquery-1.11.1.js"></script>
<script src="js/bootstrap.min.js"></script>
<script src="js/jquery.backstretch.js"></script>
<script src="js/scripts.js"></script>
</body>
</html>
```

第三步：在 src 文件夹中的 servlet 文件夹下创建并编写 RegistServlet.java 文件，验证用户名是否已经存在，若存在则重定向至注册页面，重新注册。否则，若新注册的用户信息正确，调用 addUser(userinfo) 方法，传递 userinfo 对象，向数据库新增数据，新增成功则跳转至注册成功页面，具体如示例代码 6-17 所示。

示例代码 6-17：RegistServlet.java

```
package com.xt.servlet;
import java.io.IOException;
import java.io.PrintWriter;
import javax.servlet.ServletException;
import javax.servlet.http.HttpServlet;
import javax.servlet.http.HttpServletRequest;
import javax.servlet.http.HttpServletResponse;
```

```
import com.xt.bll.UserBll;
import com.xt.dao.UserDao;
import com.xt.entity.UserInfo;
/*
 * 注册
 */
public class RegistServlet extends HttpServlet {
    private static final long serialVersionUID = 1L;
    private UserBll userbiz = null;
    @Override
    public void init() throws ServletException {
    userbiz = new UserBll();
    }
    private void check(HttpServletRequest req, HttpServletResponse resp) throws IOEx-
ception{
    String username = req.getParameter("username");
    // 检查用户名是否已存在
    boolean isUserExist = userbiz.userExists(username);
if(isUserExist){
    resp.setCharacterEncoding("utf-8");
    resp.getWriter().write("true");
    resp.getWriter().close();
    }else{
    resp.setCharacterEncoding("utf-8");
    resp.getWriter().write("false");
    resp.getWriter().close();
    }
    }
    /*
* 注册功能
*/
    private void regist(HttpServletRequest req, HttpServletResponse resp) throws IOEx-
ception{
    String username = req.getParameter("username");
    // 判断用户名是否已存在
    boolean isUserExist = userbiz.userExists(username);
if(isUserExist){
    req.setAttribute("info", " 用户名已存在 ");
```

```
        try {
        req.getRequestDispatcher("register.jsp").forward(req, resp);
        } catch (ServletException e) {
        // TODO Auto-generated catch block
        e.printStackTrace();
        }
        }else{
        String password = req.getParameter("password");// 使用 getParameter 获取属性值
        String email = req.getParameter("email");
        UserInfo userinfo = new UserInfo();
        userinfo.setUsername(username);// 将 username 变量值赋给 userinfo 对象中的 username
        userinfo.setPassword(password);
        userinfo.setEmail(email);
        if(!userbiz.addUser(userinfo)){
        resp.setCharacterEncoding("utf-8");
        PrintWriter pw = resp.getWriter();// 使用
        pw.println("<script type=\"text/javascript\">");
        pw.println("alert(\" 注册失败 \");");
        pw.println("open(\"login.jsp\",\"_self\");");
        pw.println("</script>");
        return;
        }
        resp.sendRedirect("register_success.html");
        }
        }
        @Override
        protected void doPost(HttpServletRequest req, HttpServletResponse resp)
        throws ServletException, IOException {
        regist(req, resp);
        }
    }
```

第四步：编写 UserService 文件用来验证用户是否存在及用户是否添加成功，返回值为 boolean 类型，具体如示例代码 6-18 所示。

示例代码 6-18：UserService.java

```
package com.xt.service;
import java.util.HashMap;
import java.util.List;
```

```
import com.xt.dao.UserDao;
import com.xt.entity.UserInfo;
public class UserService {
    private UserDao userdao = new UserDao();
    public boolean addUser(UserInfo userinfo) {
    int row = userdao.insert(userinfo);
    return row>0?true:false;
    }
    /*
* 执行登录请求操作
*/
    public int checkLogin(String username, String password) {
    UserInfo userInfo = new UserInfo();
    userInfo.setUsername(username);
    userInfo.setPassword(password);
    List<HashMap> list = userdao.query(userInfo);
    return list.size()>0?(int)list.get(0).get("uid"):-1;
    }
    /*
* 查看用户是否存在
*/
    public boolean userExists(String username) {
    List<HashMap> list = userdao.userExists(username);
    return list.size()>0?true:false;
    }
}
```

第五步：编写 UserDao 文件，功能为编写 SQL 语句、设置数据，具体如示例代码 6-19 所示。

示例代码 6-19：UserDao.java

```
package com.xt.dao;

import java.util.ArrayList;
import java.util.HashMap;
import java.util.List;

import com.xt.entity.UserInfo;
```

```
public class UserDao extends BaseDao {
    public int insert(UserInfo userinfo) {

    String sql = "INSERT INTO userinfo VALUES(NULL,?,?,?)";// 添加用户，对应添加
的字段为自增 id 字段、用户名、密码和邮箱，其中 NULL 为占据自增 id 字段位置，在新增
时会自动添加 id

    Object[]   params   =   {userinfo.getUsername(),userinfo.getPassword(),userinfo.getE-
mail()};

    return doCUD(sql,params);// 调用 doCUD 方法传递 sql 语句与数据
    }
    public List<HashMap> userExists(String username){
    // 查询 userinfo 表是否存在用户
    String sql = "SELECT * FROM userinfo WHERE username =?";
    Object[] params = {username};

    return doSelect(sql,params);
    }
}
```

第六步：编写公共方法 BaseDao，在已有的 BaseDao 文件中添加 doCUD(String sql, Object[] params) 方法，完成对应的新增、修改、删除，具体如示例代码 6-20 所示。

示例代码 6-20：BaseDao.java

```
public class BaseDao {
    /*
* 执行新增、修改、删除的方法
*/
    public int doCUD(String sql, Object[] params) {
    int rows = 0;
    try {
    openconnection();
    ps = conn.prepareStatement(sql);// 传递 sql 语句
    for (int i = 0; i < params.length; i++) {
    ps.setObject(i + 1, params[i]);// 获取用户填写的数据
    }
    rows = ps.executeUpdate();// 调用方法对数据库进行操作
    System.out.println(" 本次操作共更新数据 " + rows + " 条 ");
```

```
        } catch (Exception e) {
        e.printStackTrace();
        } finally {

        closeResource();

        }

        return rows;
    }
    }
```

第七步：配置 web.xml 中 Servlet 映射关系文件，具体如示例代码 6-21 所示。

示例代码 6-21：web.xml

```
<servlet>
  <description> 注册组件 </description>
  <servlet-name>regist</servlet-name>
  <servlet-class>com.xt.servlet.RegistServlet</servlet-class>
</servlet>
<servlet-mapping>
  <servlet-name>regist</servlet-name>
  <url-pattern>/regist</url-pattern>
</servlet-mapping>
```

完成上述操作后，网上书城项目注册模块就完成了，效果如图 6-11 至图 6-13 所示。

通过对 Servlet 知识的学习和制作网上书城项目中的注册模块，了解 Servlet 的基本概念和原理，掌握 Servlet 的使用方法，掌握过滤器和监听器的使用方法，能够通过使用 Servlet 处理响应完成注册模块的功能。

destroy	销毁	generic	普通的
listener	聆听者	filter	过滤器

1. 选择题

1）当访问一个 Servlet 时，以下 Servlet 中的（　　）方法先被执行。

A.destroy()　　B.doGet()　　C.service()　　D.init()

2）Servlet 接收请求时，会调用（　　）方法。

A .service()　　B.doGet()　　C.doPost()　　D.init()

3）Servlet 中，HttpServletResponse 的（　　）方法用来把一个 Http 请求重定向到另外 URL。

A.sendURL()　　B.redirectURL()

C.sendRedirect()　　D.redirectResponse()

4）一个 Servlet 的生命周期不包括（　　）。

A.init() 方法　　B.invalidate() 方法　　C.service() 方法　　D.destroy() 方法

5）假设在 myServlet 应用中有一个 MyServlet 类，在 web.xml 文件中对其进行如下配置：

<servlet>
<servlet-name> myservlet </servlet-name>
<servlet-class> com.xt.MyServlet </servlet -class>
</servlet>
< servlet-mapping>
<servlet -name> myservlet </servlet-name>
<servlet-pattern> /welcome </url-pattern>
</servlet-mapping>

则以下选项可以访问到 MyServlet 的是（　　）。

A.http://localhost:8080/MyServlet

B.http://localhost:8080/myservlet

C.http://localhost:8080/com/xt/MyServlet

D.http://localhost:8080/welcome

2. 程序题

本实例共包括 enter.jsp、show.jsp 和 ServletDemo.java 3个文件，有如下几个要求。

1）在 enter.jsp 文件中有一个表单、文本输入框以及提交按钮，在单击“提交”按钮时发送至 ServletDemo 文件中。

2）在 ServletDemo 文件中，根据 enter.jsp 页面输入的信息进行判断，若发送值为“JSP”则使用转发跳转值 show.jsp 页面；当用户输入为其他字符时，使用重定向跳转至 show.jsp 页面。

3）对于 show.jsp 页面，根据两种跳转的区别，分析两种跳转的不同。

第七章　网上书城项目应用MVC设计模式

通过学习 MVC 的设计模式，了解 MVC 模式在 Web 应用中的优势，熟悉 MVC 模式的具体组成，具有运用学习的相关知识，编写网上书城项目主页数据显示和搜索功能以及下单功能的能力。在任务实现过程中：

- 了解 MVC 设计模式；
- 掌握 MVC 设计模式中 Model 的实现；
- 掌握 MVC 设计模式中 View 的实现；
- 掌握 MVC 设计模式中 Control 的实现。

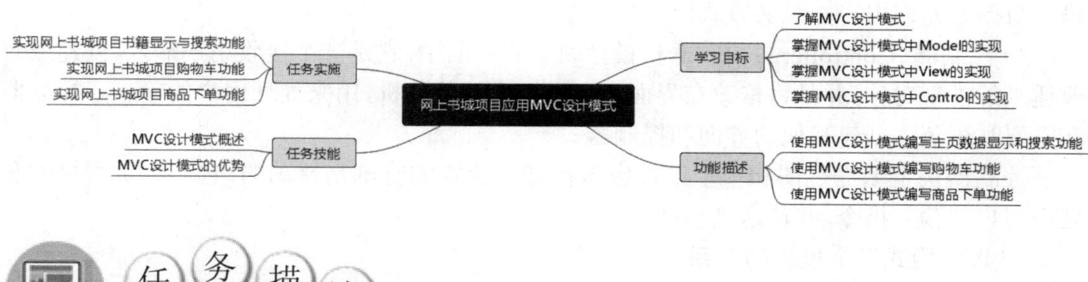

任务描述

【情境导入】

在 JavaWeb 开发过程中，一个完整的模块可能包括业务逻辑代码、数据库操作代码、

HTML 表示层代码等，为了提高 Web 程序的可扩展性、可复用性、可维护性，我们引入一种开发模式来分层开发程序，以达到上述优化目的。

【功能描述】

- 使用 MVC 设计模式编写主页数据显示和搜索功能。
- 使用 MVC 设计模式编写购物车功能。
- 使用 MVC 设计模式编写商品下单功能。

技能点一　MVC 设计模式概述

在 Web 程序不断发展的过程中，程序执行的业务逻辑逐渐多样化，网站的视图页面也日益复杂化，为了降低不同功能代码之间的耦合度，提高系统组件的重用性，以及更加便于小组之间的合作开发，大部分的 JavaWeb 应用都采用了 MVC 模式来进行分层开发。

MVC 模式的全称为 Model-View-Controller（模型 - 视图 - 控制器）模式，MVC 将程序以功能为区分，分割为 3 个模块，每个模块的作用如下。

1）模型（Model）：模型包含完成任务所需要的所有行为和数据，一般由许多类组成，并且使用面向对象的技术实现在数据库中存取数据的功能。

2）视图（View）：视图是用户看到并与之交互的界面，一个界面就是一个程序的可视化元素。视图的数据由模型提供，它并不控制数据或提供除显示外的其他行为，只是作为一种输出数据并允许用户操纵的方式。

3）控制器（Controller）：控制器将模型映射到界面中，它是一个接收用户输入、创建或修改适当的模型对象，并且将修改在界面中体现出来的状态机，用来处理用户的输入。控制器在需要时还负责创建其他的界面和控制器。

控制器决定哪些视图和模型组件应该在某个给定的时刻是活动的，它一直负责接收和处理用户的输入指令，并传送到模型。

1.MVC 模式三个模块的关系

MVC 模式三个模块之间的关系如图 7-1 所示。

控制器控制返回的视图及模型，调用被选择的模型和视图以继续执行流程。一个应用程序可能有多个控制器，每个控制器负责应用的某个特定领域。通过协调用户请求的回答，控制器管理着全部的应用流程。

模型存储应用状态和一些业务为控制器和视图提供了统一的接口。视图从模型中读取数据，并使用这些数据来生成应答。MVC 关系图的分工和协作如表 7-1 所示。

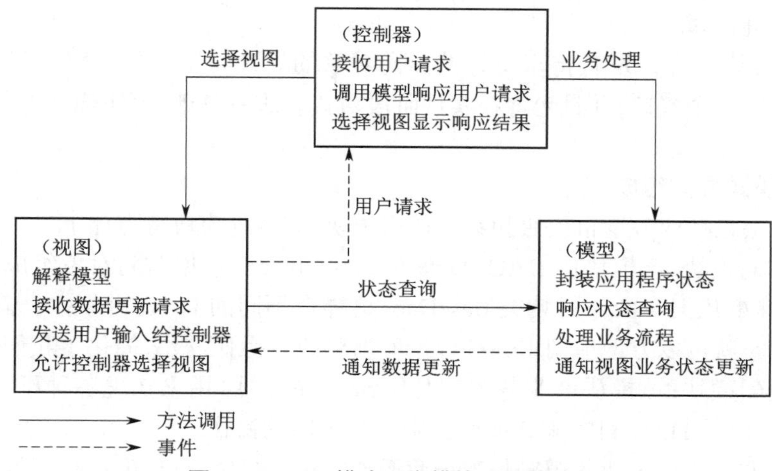

图 7-1　MVC 模式三个模块之间的关系

表 7-1　MVC 关系图的分工和协作说明

	模型 M	视图 V	控制器 C
分工	抽象系统应用的功能； 封装系统的状态； 提供使用系统功能的方法和路径； 管理数据的存储和一致性； 当数据发生变化时通知相关部分	抽象数据的表达式； 表示针对用户的数据； 维护与 Model 数据的一致性	抽象用户和系统的事件的语意映射； 把用户输入翻译为系统事件； 根据用户输入和上下文情况选择合适的数据
协作	当改变系统数据时通知 View； 能够被 View 检索数据； 提供对 Controller 的操作	把 Model 表征给用户； 当数据被相关 Model 改变时更新表示的数据； 把用户输入提交给 Controller	把用户输入转化成对 Model 的系统行为； 根据用户输入和 Model 的动作结果选择合适的 View

MVC 方法的实现在很大程度上是基于一个事件驱动环境，由用户通过使用界面来控制应用的流程。Smalltalk 是最早集成 GUI 的面向对象语言，MVC 首先是在 Smalltalk 中实现的，由于这个语言的创建，使得 MVC 设计方法的想法得以实现。因此，MVC 最先是在 GUI 应用程序上得到应用。

MVC 通过以下四种方式消除与用户接口和面向对象的设计有关的绝大部分困难。

1）控制器通过一个状态机跟踪和处理面向操作的用户事件。这允许控制器在必要时创建和破坏来自模型的对象，并且将面向操作的拓扑结构与面向对象的设计隔离开来，这个隔离有助于防止面向对象的设计走向反面。

2）MVC 将用户接口与面向对象模型分开，允许同样的模型不用修改直接使用许多不同的界面方式。如果模型更新由控制器完成，那么界面就可以多次使用。

3）MVC 允许应用的用户接口进行大的变化而不影响模型，每个用户接口的变化将只需要对控制器进行修改。

4）MVC 设计允许开发者将一个好的面向对象的设计与用户接口隔离开来，允许在同样的模型中使用多个接口，并且允许在实现阶段对接口进行修改，而不需要对相应的模型进行修改。

2 .MVC 模式具体组成

1）Model 层包括业务逻辑模块和数据访问模块，这两种模块特点如下。

①数据访问模块：该模块由 DAO（Data Access Object）数据访问对象组成，本书项目中 DAO 对象需继承数据库操作封装类 BaseDao，这样在不同的 DAO 对象中只需要编写 SQL 语句和参数对象就可以方便地调用各种数据库增删改查操作，极大简化了我们的开发过程。

②业务逻辑模块：该模块由业务逻辑对象 Service 组成，用来实现各种不同业务的逻辑操作，并调用对应的 DAO 对象来完成业务中对数据库的操作。

这种分层方式解耦了业务逻辑与数据库操作方法，提高了程序的可扩展性和可维护性。

2）View 层由不同的 JSP 页面组成，用户通过 View 层浏览系统数据，并执行相关操作与 Web 应用进行交互。

3）Controller 层由响应不同请求的 Servlet 对象组成，Controller 控制 Model 层与 View 层，同步 View 层和 Model 层的数据，在 Web 应用中起到调度的作用。

技能点二　MVC 设计模式的优势

Web 的应用几乎完全是用户驱动的，所以必须为程序提供业务逻辑与显示逻辑。在使用 JSP 开发 Web 时，我们把显示逻辑（HTML）嵌在应用的业务逻辑中，开发人员会使用一系列的 JSP 页面来实现应用的业务逻辑，同时还要向用户显示界面，这称为 Model 1 体系，结构如图 7-2 所示。

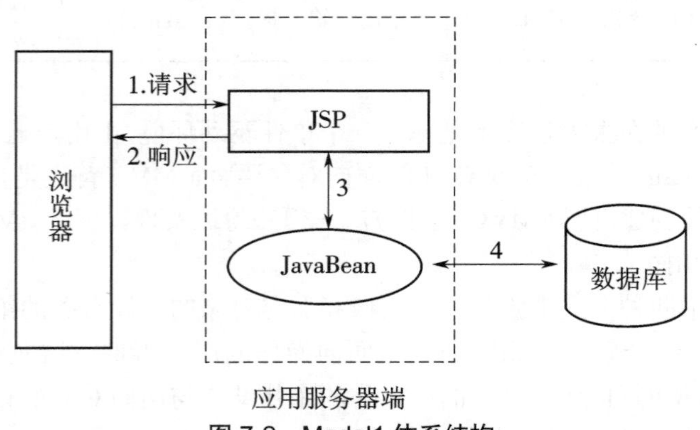

图 7-2　Model1 体系结构

当用户请求一个 Web 页面时，Tomcat 执行其中的逻辑和包含的页面，这种执行可能包括从数据库或其他函数获取数据来满足业务逻辑。JavaBean 提供了 JSP 页面中的数据表示，在页面业务逻辑中包含一些生成 HTML 的代码，通过生成这些 HTML 显示给用户，作为处理结果，其会构建最终的页面，显示给用户。

对于一个小型 Web 应用程序来说，它的业务逻辑很有限，这种体系结构相当合适。但在一个较复杂的应用中，业务逻辑更有深度，所需的显示逻辑也比较复杂，此时如果仍采用 Model 1 架构模式，会导致程序混乱且无法维护。以下为该模式的缺点。

1）代码重复：完成某些业务规则或影响显示内容的代码通常会重复。由于某些逻辑会多次出现，修改起来相当困难。

2）可维护性差：当业务逻辑与显示逻辑混合在一起，修改不当会导致其他问题的发生，增加维护、二次开发的困难。

3）可扩展性差：应用程序中业务逻辑如果有修改的话，可能要求完成大规模的重构。在一个 Web 应用程序开发中，一个简单的修改就会付出很大的代价。

4）可测试性差：若测试过程中发生问题，不容易定位到关键问题部分。

为弥补 Model1 架构模式的缺点，开发人员创建了一个使用 Servlet、JSP 以及 JavaBean 的 Model2 体系结构，它以 MVC 体系结构为基础，需要用一个 Servlet 作为控制器，接收来自用户的需求，影响模型中的修改，并向客户提供视图。这种优化的体系结构如图 7-3 所示。

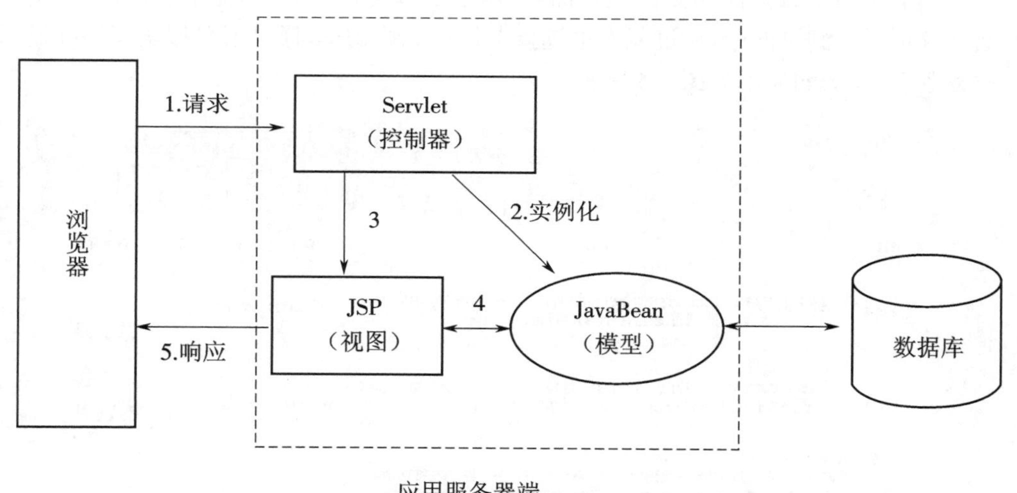

图 7-3　Model2 体系结构

这个体系结构中实现的视图仍然使用 JSP 页面，但是其中包含的逻辑只与用户显示界面有关，而不再实现业务逻辑。模型层封装在 Java 对象中，对象并不关心会如何显示。该体系的执行步骤如下。

第一步：用户请求一个 Servlet 的 URL。

第二步：控制器接收这个请求，并基于请求确定要完成的工作。控制器在模型中执行调用来具体完成所需的业务逻辑。

第三步：控制器指示模型层提供一个数据对象，模型可能要访问数据库来提供这些数据。

第四步：控制器得到一个数据对象，以便在视图中显示，控制器还要确定适当的视图提供给用户。通过使用请求分派器，这个控制器（Servlet）可以为所选择的视图（JSP）提供数据对象。

第五步：视图有了所提供的数据对象，它会根据其显示逻辑来提供数据的显示。

第六步：作为这个处理的结果，所生成的 HTML 会作为响应发回给用户。

通过上面的分析，可以从高层次的角度将一个应用的对象分为三类：① JavaBean 包含商业规则和数据；② Servlet 就是接收请求，控制商业对象去完成请求；③ JSP 是用户显示部分，管理整体页面的样式、风格和内容等。

通常当系统发布后，View 对象是由美工、HTML/JSP 设计人员或者系统管理员来负责管理的。Controller 对象由应用开发人员开发实施，Model 业务逻辑对象则由开发人员、领域专家和数据库管理员共同完成。

1.MVC 实现书籍显示与搜索功能

网上书城项目登录后跳转到的主页面中，由于书本类目过多，不便于显示，需要实现查询书籍并分页显示到页面中，通过复选框选择书籍加入购物车，通过书名搜索书籍并显示这三种业务逻辑，页面的显示如图 7-4 所示。

图 7-4　网上书城项目书籍列表

具体步骤如下。

（1）书籍显示 Model 层

书籍显示功能的 Model 层需要实现根据页数查询书籍和根据书名查询书籍。在

service 项目文件中创建 BookService.java 类，在 dao 项目文件中创建 BookDao.java 类，来编写这两项功能相关的业务逻辑代码和数据访问代码，如图 7-5 所示。

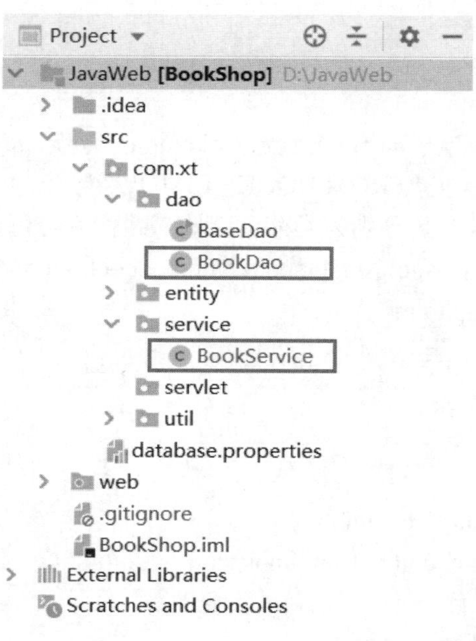

图 7-5　创建 Model 层

编写 BookDao 类，该类继承自数据库操作封装类 BaseDao，在类中编写根据页数查询书籍的 findBooks 方法、根据书名查询书籍的 findBookByName 方法以及查询数据库中书籍总数的 count 方法，书籍总数用于确定 View 分页的总页数，具体如示例代码 7-1 所示。

示例代码 7-1：BookDao.java

```
package com.xt.dao;
import java.util.HashMap;
import java.util.List;
import com.xt.entity.Book;
/**
 * 书籍数据访问模块
 */
public class BookDao extends BaseDao {
/*
* 根据书名查询书籍
*/
    public List<HashMap> findBookByName(String BookName) {
    String sql = "select * from books where bookname =?";
    Object[] params = {BookName};
```

```
        return doSelect(sql,params);
        }
    /*
    * 根据页数查询书籍
    */
        public List<HashMap> findBooks(int bookNumber,int page_NO){
        String sql = "SELECT * FROM BOOKS LIMIT ?,?";
         // 为占位符赋值，参数分别为当前页起始书籍位置与每页数据条数
        Object[] params = {(page_NO-1)*bookNumber,bookNumber};
        return doSelect(sql,params);
        }
    /*
    * 查询书籍总数
    */
        public List<HashMap> count() {
        String sql = "select count(*) from books";
        Object[] params = {};
        return doSelect(sql,params);
        }
    }
```

编写 BookService 类来实现书籍的业务逻辑，在类中实例化 BookDao 对象来执行业务中的数据处理操作，具体如示例代码 7-2 所示。

示例代码 7-2：BookService.java

```
package com.xt.Service;
import java.math.BigDecimal;
import java.util.ArrayList;
import java.util.HashMap;
import java.util.List;
import com.xt.dao.BookDao;
import com.xt.entity.Book;
/*
书籍业务逻辑模块
*/
public class BookService {
   // 实例化 bookdao 对象
    BookDao bookdao = new BookDao();
/*
```

```
* 根据当前页数获取图书
*/
    public List<Book> findBooks(int page_books, int page_NO) {
    List<HashMap> mapList = bookdao.findBooks(page_books,page_NO);
     // 将查询后得到的 Map 集合转化为 Book 对象集合
    List<Book> bookList = new ArrayList<>();
    for(HashMap bookmap : mapList){
    Book book = new Book();
    book.setBid((int)bookmap.get("bid"));
    book.setBookname((String)bookmap.get("bookname"));
    book.setPrice((BigDecimal)bookmap.get("price"));
    book.setStock((int)bookmap.get("stock"));
    book.setImage((String)bookmap.get("image"));
    book.setBooknumber((String)bookmap.get("booknumber"));
    book.setIntroduction((String)bookmap.get("introduction"));
    bookList.add(book);
    }
    return bookList;
    }
/*
* 按照书名查找图书
*/
    public List findBookByName(String bookName) {
    return bookdao.findBookByName(bookName);
    }
/*
* 返回图书数量
*/
    public Long count() {
      // 从查询后得到的 Map 集合中获取书籍总数
List<HashMap> mapList = bookdao.count();
Long total = (Long)mapList.get(0).get("count(*)");
return total;
}
}
```

（2）书籍显示与书籍搜索 Controller 层

在 Servlet 包中创建 ShowBooksServlet.java 类，该 Servlet 用于渲染主页面 main.jsp，使书籍列表动态显示在页面上，具体如示例代码 7-3 所示。

示例代码 7-3：ShowBooksServlet.java

```
package com.xt.servlet;
import java.io.IOException;
import java.util.List;
import javax.servlet.ServletException;
import javax.servlet.http.HttpServlet;
import javax.servlet.http.HttpServletRequest;
import javax.servlet.http.HttpServletResponse;
import com.xt.Service.BookService;
import com.xt.util.PageTools;
/**
 * 显示全部图书
 */
public class ShowBooksServlet extends HttpServlet {
    private BookService bookSerivce = null;
    @Override
    public void init() throws ServletException {
     //Servlet 初始化时实例化 bookSerivce 对象
    bookSerivce = new BookService();
    }
    @Override
    protected void service(HttpServletRequest req, HttpServletResponse resp)
    throws ServletException, IOException {
      // 获取当前页面的页数
    String NO_str = req.getParameter("current_books_NO");
      // 将当前页数转为 int 对象，如果是首次请求页面，该对象为空时，则为其赋值 1
    int NO = NO_str==null?1:Integer.valueOf(NO_str);
      // 查询书籍总数
    int total_books = bookSerivce.count().intValue();
      // 获取分页的每页数据条数
    int book_num = PageTools.book_num;
      // 根据书籍总数和每页数据条数计算总页数
        int total_page = total_books%book_num==0?total_books/book_num:total_books/
book_num+1;
      // 根据当前页数搜索书籍数据
    List books = bookSerivce.findBooks(PageTools.book_num, NO);
      // 将书籍信息 books、当前页数 NO、总页数 total_page 存入 request 对象中
    req.setAttribute("books", books);
```

```
        req.setAttribute("current_books_NO", NO);
        req.setAttribute("total_books_page", total_page);
        req.getRequestDispatcher("main.jsp").forward(req, resp);
        }
    }
```

在 Servlet 包中创建 SearchServlet.java 类，该 Servlet 用于响应书籍搜索框，使查询结果显示到 main.jsp 页面中，具体如示例代码 7-4 所示。

示例代码 7-4：SearchServlet.java

```
package com.xt.servlet;
import java.io.IOException;
import java.util.List;
import javax.servlet.ServletException;
import javax.servlet.http.HttpServlet;
import javax.servlet.http.HttpServletRequest;
import javax.servlet.http.HttpServletResponse;
import com.xt.Service.BookService;
/**
 * 书籍搜索框
 */
public class SearchServlet extends HttpServlet {
    private BookService bookSerivce = null;
//Servlet 初始化时实例化 bookSerivce 类
    @Override
    public void init() throws ServletException {

    bookSerivce = new BookService();
    }
/**
 * 获取请求中的书籍名称并调用查询业务
 * 将查询结果返回 JSP 页面
 */
    @Override
    protected void service(HttpServletRequest req, HttpServletResponse resp)

    throws ServletException, IOException {
    String keywords = req.getParameter("keywords");
```

```
        List books =  bookSerivce.findBookByName(keywords);

        req.setAttribute("books", books);

        req.getRequestDispatcher("main.jsp").forward(req, resp);
        }
    }
```

（3）书籍显示与书籍搜索 View 层

主页面的 View 层 main.jsp 页面代码如示例代码 7-5 所示。该页面使用 EL 表达式将 Controller 层通过 Request 对象传到页面中的 Book 集合动态显示出来，并可以通过复选框选中书籍，将书籍添加到购物车中。

示例代码 7-5：main.jsp

```
<%@page language="java" import="java.util.*" contentType="text/html; charset=utf-8"
isELIgnored="false"%>
<%@page import="com.xt.util.PageTools"%>
<%@page import="com.xt.entity.Book"%>
<%@taglib uri="http://java.sun.com/jstl/core_rt" prefix="c" %>
<jsp:include page="elements/index_head.jsp"/>// 引入头部文件
<link rel="stylesheet" href="css/bootstrap.css">
<script src="js/jquery.js"></script>
<script src="js/bootstrap.js" ></script>
<link rel="stylesheet" href="css/style4.css">
<link rel="stylesheet" href="css/icomoon.css">
<link rel="stylesheet" href="css/style.default.css" id="theme-stylesheet">
<!-- Custom stylesheet - for your changes-->
<link rel="stylesheet" href="css/custom.css">
<%// 页面分页逻辑
    int NO = 0;
    int total_page = 0;
    List books = (List)request.getAttribute("books");
    if(request.getAttribute("current_books_NO") != null){

    NO = (Integer)request.getAttribute("current_books_NO");

    total_page = (Integer)request.getAttribute("total_books_page");
    }
%>
```

```
<body>
<jsp:include page="elements/main_menu.jsp"/>// 引入导航栏
<div id="fh5co-pricing" class="fh5co-bg-section">
<div class="container">
<div class="row animate-box">
<div class="col-md-12  text-center fh5co-heading">
<h2> 书籍列表 </h2>
</div>
</div>
<div class="row">// 书籍信息显示,通过表格形式进行显示
<div id="basket1" class="col-lg-12">
<div class="box">
<form form method="post" name="shoping" action="cart">
<div class="table-responsive">
<table class="table">
<thead>
<tr class="title">
<th class="checker"></th>
<th> 书名 </th>
<th> 简介 </th>
<th> 编号 </th>
<th class="price"> 价格 </th>
<th class="store"> 库存 </th>
<th class="view"> 图片预览 </th>
</tr>
</thead>
<tbody>
<c:forEach var="book" items="${books}">
<tr>
<td><input type="checkbox" name="bookId" id="bookId" value="${book.bid}"/></td>
<td class="title">${book.bookname}</td>
<input type="hidden" name="title" value="${book.bid}:${book.bookname}"/>
<td>${book.introduction}</td>
<input type="hidden" name="introduction"  value="${book.bid}:${book.introduction}"/>
<td>${book.booknumber}</td>
<input type="hidden" name="booknumber" value="${book.bid}:${book.booknumber}"/>
<td> ￥${book.price}</td>
<input type="hidden" name="price" value="${book.bid}:${book.price}"/>
```

```
<td>${book.stock}</td>
<input type="hidden" name="stock" value="${book.bid}:${book.stock}"/>
<td class="thumb"><img src="${book.image}" width="100px" height="100px"/></td>
<input type="hidden" name="image" value="${book.bid}:${book.image}"/>
</tr>
</c:forEach>
</tbody>
</table>
<%if (request.getAttribute("current_books_NO") != null) { %>// 编写分页部分
<nav aria-label="Page navigation">
<ul class="pagination">
<li>
<a href="books"> 首页 </a>
</li>
<%for (int i = 1; i <= total_page; i++) { %>// 根据总共的页数进行分页

<%if (i == NO) { %>
<li><a href="current"><%=i %>
</a></li>
<%
continue;
}
%>
<li><a href="books?current_books_NO=<%=i %>"><%=i %>
</a></li>
<%} %>
<li><a href="books?current_books_NO=<%=total_page %>"> 页尾 </a></li>
</ul>
</nav>
<%} %>
</div>
<div class="box-footer d-flex justify-content-between flex-column flex-lg-row">
<div class="button">
<input class="input-btn add" type="submit" name="submit" value=" 添加到购物车 " on-
click="cart()"/>
</div>
</div>
</form>
```

```
</div>
</div>
</div>
</div>
</div>
</body>
<jsp:include page="elements/main_bottom.html"/>
<script type="text/javascript">
  function cart() {
    var input = document.getElementsByTagName('input');
    var countCheckBox = 0;
    var countChecked = 0;
    for (var i = 0; i < input.length; i++) {
      if (input[i].checked === true) {
        countChecked++;// 获取 checkbox 被勾上的数量
      }
    }
    if (countChecked == 0) {
      alert(" 请至少勾选一件商品 ");
      window.location.href = "main.jsp";
    }
}
</script>
```

2.MVC 实现添加购物车功能

添加购物车的业务逻辑是获取请求对象里以参数形式传到 Servlet 中的书籍 bid 集合，将选中书籍的 bid 以及购买数目以键值对的形式保存在 Session 对象中。该逻辑只涉及 Controller 层和 View 层的交互，具体步骤如下。

第一步：在 Servlet 包中创建 CartServlet.java 类来实现该逻辑，具体如示例代码 7-6 所示。

示例代码 7-6：CartServlet.java

```
package com.xt.servlet;
import java.io.IOException;
import java.util.HashMap;
import javax.servlet.ServletException;
import javax.servlet.http.HttpServlet;
import javax.servlet.http.HttpServletRequest;
import javax.servlet.http.HttpServletResponse;
```

```
/**
 * 将书籍添加到购物车
 */
public class CartServlet extends HttpServlet {
    @Override
    protected void doPost(HttpServletRequest req, HttpServletResponse resp)

    throws ServletException, IOException {

    req.setCharacterEncoding("UTF-8");

    resp.setCharacterEncoding("utf-8");

    String[] bids = req.getParameterValues("bookId");

    // 获取购物车内容，没有则初始化一个列表

    HashMap shoppingCart = (HashMap) req.getSession().getAttribute("shoppingCart");

    if(shoppingCart == null)

    shoppingCart = new HashMap();

    // 获取复选框选中的值

    for(int i = 0; i < bids.length; i++){

    Object currentCount = shoppingCart.get(bids[i]);

    // 查看当前图书是否已经存在于购物车中

    if(currentCount!=null){
    shoppingCart.put(bids[i],(int)currentCount+1);

    }else{

    shoppingCart.put(bids[i],1);
```

```
        }

    }

    req.getSession().setAttribute("shoppingCart", shoppingCart );

    req.getSession().setAttribute("CartBookNum",shoppingCart.size());

    resp.sendRedirect("shopping-success.jsp");
    }
}
```

第二步：编写 web.xml 文件中的 Servlet 映射关系，具体如示例代码 7-7 所示。

示例代码 7-7：Servlet 映射配置

```
<servlet>
  <description> 显示书籍组件 </description>
  <servlet-name>showBooks</servlet-name>
  <servlet-class>com.xt.servlet.ShowBooksServlet</servlet-class>
</servlet>
<servlet-mapping>
  <servlet-name>showBooks</servlet-name>
  <url-pattern>/books</url-pattern>
</servlet-mapping>
<servlet>
  <description> 查找图书 </description>
  <servlet-name>searchBooks</servlet-name>
  <servlet-class>com.xt.servlet.SearchServlet</servlet-class></servlet>
<servlet-mapping>
  <servlet-name>searchBooks</servlet-name>
  <url-pattern>/search</url-pattern>
</servlet-mapping>
<servlet>
  <description> 添加购物车 </description>
  <servlet-name>addToCart</servlet-name>
  <servlet-class>com.xt.servlet.CartServlet</servlet-class>
</servlet>
<servlet-mapping>
  <servlet-name>addToCart</servlet-name>
```

```
    <url-pattern>/cart</url-pattern>
</servlet-mapping>
```

3.MVC 实现下单功能

运用 MVC 模式完成网上书城项目中的从购物车下单功能。该功能包括两个页面，分别为购物车页面（图 7-6）和订单页面（图 7-7）。

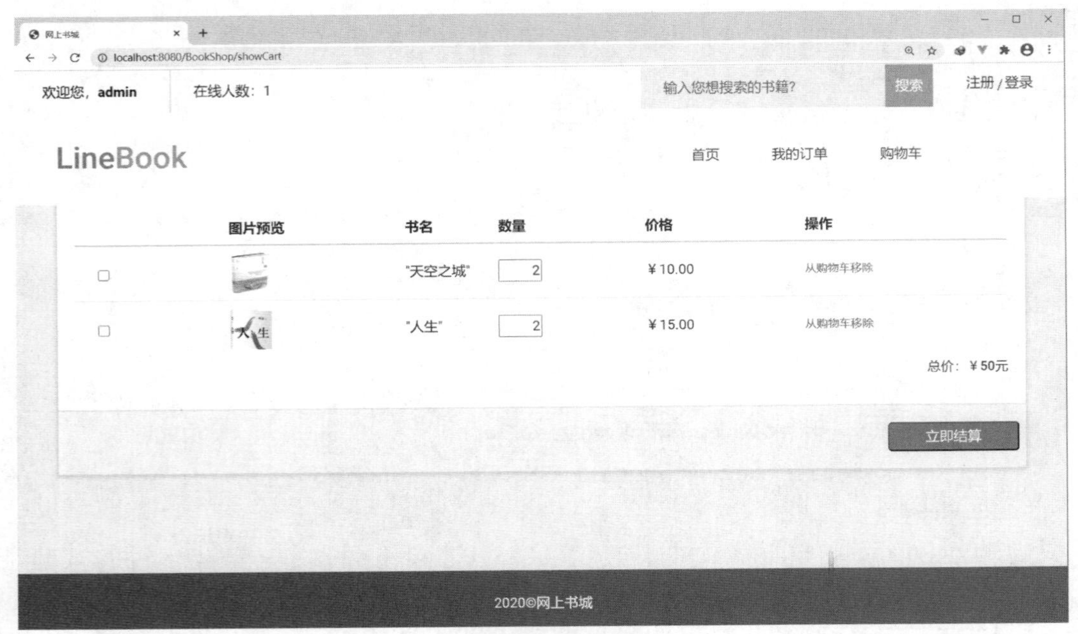

图 7-6　购物车页面

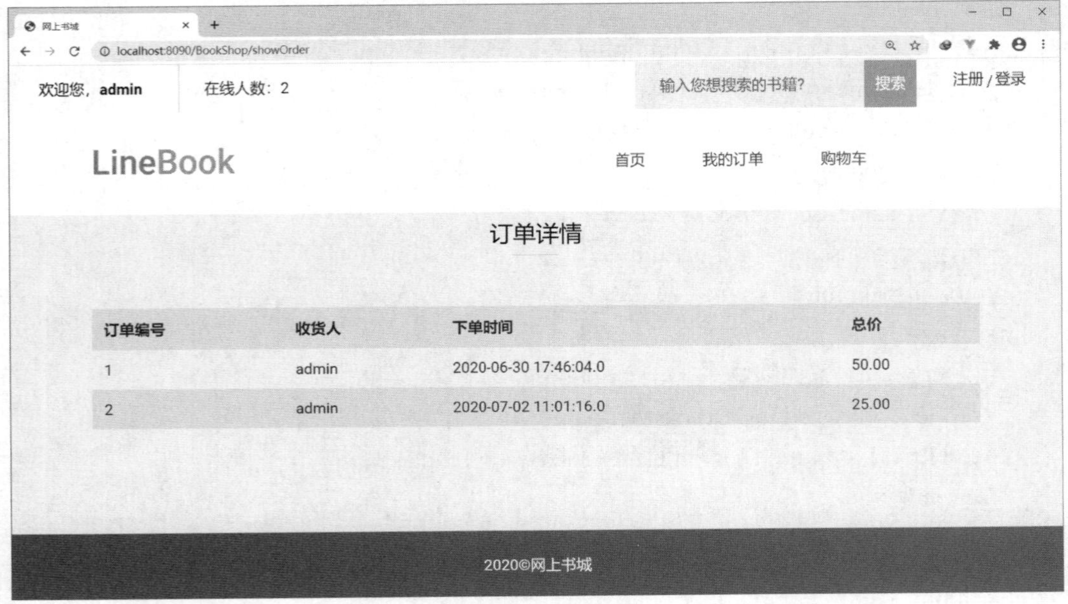

图 7-7　订单页面

下单功能包括：①在购物车页面中显示并修改购买的书籍；②在购物车结算下单；③结算下单成功后跳转，其中需要重点注意的业务逻辑有以下三点：

1）在购物车 shopping.jsp 页面中显示购物车中添加书籍的信息以及订单的总额；

2）在购物车 shopping.jsp 页面中可执行修改购买数量、移除购物车中书籍以及结算下单功能；

3）成功结算后跳转到用户订单显示页面。

实现该功能具体步骤如下。

（1）下单模块 Model 层

下单功能的 Model 层需要实现根据书籍 bid 查询书籍信息的功能、结算下单功能以及查询用户订单功能，创建 OrderService.java 和 OrderDao.java 来编写结算下单和显示用户订单相关的业务逻辑代码和数据访问代码，创建 BookService.java 和 BookDao.java 来编写查询书籍信息的业务逻辑代码和数据访问代码。

根据 bid 查询书籍业务逻辑编列流程，在 BookDao.java 文件中编写查询方法 findBookByBID，具体如示例代码 7-8 所示。

示例代码 7-8：findBookByBID 方法

```
public List<HashMap> findBookByBID(int bid){
    String sql = "select * from books where bid = ? ";
    Object[] params = {bid};
    return doSelect(sql,params);
}
```

在 BookService 类中编写 BookService 方法，具体如示例代码 7-9 所示。

示例代码 7-9：BookService.java

```
public Book findBookByBID(int bid){
    HashMap bookmap = bookdao.findBookByBID(bid).get(0);
// 将查询后得到的 Map 对象转化为 Book 对象
    Book book = new Book();
    book.setBid((int)bookmap.get("bid"));
    book.setPrice((BigDecimal) bookmap.get("price"));
    book.setStock((int)bookmap.get("stock"));
    book.setImage((String)bookmap.get("image"));
    book.setBookname((String) bookmap.get("bookname"));
    return book;
}
```

结算功能的主要逻辑是先在 orders 表中创建订单信息，再将订单中的商品信息写入 items 表中，最后在 book 表中更新数据，减去购买书籍的库存数，这三步必须同时成功，为保证数据库中数据的可靠性和一致性，需要使用 JDBC 事务功能来进行该流程。

在 dao 包中创建 OrderDao 类，并在该类中编写下单方法 placeOrder，由于在事务进行过程中，无法使用 getGeneratedKeys() 获得 Insert 方法的自增主键，故需要先使用查询语句获得 orders 表的最新 oid 值。

获取 orders 表主键值后，即可开启事务，执行向数据库中新增 orders 条目和 items 条目并更新 book 表中书籍库存。该方法需要传入 2 个参数，分别是订单 JavaBean 对象 order 以及订单商品 JavaBean 对象集合 itemList，方法执行成功则返回 1，如执行失败或事务回滚则返回 -1。

在 OrderDao 类中创建 findOrderByUID 方法来通过用户 uid 查询用户所有订单，具体如示例代码 7-10 所示。

示例代码 7-10：OrderDao.java

```
package com.xt.dao;
import java.sql.SQLException;
import java.sql.Statement;
import java.util.HashMap;
import java.util.List;
import com.xt.entity.Item;
import com.xt.entity.Order;
/**
 * 订单数据访问模块
 */
public class OrderDao extends BaseDao {
/**
 * 根据用户 uid 查询用户所有订单
 */
    public List<HashMap> findOrderByUID(int uid){
    String sql="SELECT oid FROM orders WHERE uid=?";
    Object[] params = {uid};
    return doSelect(sql,params);
    }
    /**
 * 订单结算
*/
    public int placeOrder(List<Item> itemList, Order order){
    // 获取当前 orders 表单自增主键
    String sql2 = "SELECT oid FROM orders ORDER BY oid DESC LIMIT 1";
    List<HashMap> oidL = doSelect(sql2,new Object[]{});
    int oid = 1;
    if(oidL.size()!=0){
    oid = (int)oidL.get(0).get("oid")+1;
```

```
        }
        // 开始新增 order 和 items
        openconnection();
        try {
    // 开启事务
        conn.setAutoCommit(false);
        // 执行新增 order
        String sql = "INSERT INTO orders VALUES(?,?,?,?)";
        ps = conn.prepareStatement(sql, Statement.RETURN_GENERATED_KEYS);
        Object[] params = {oid,order.getUid(),order.getTotal_price(),order.getCreateDate()};
        for (int i = 0; i < params.length; i++) {
        ps.setObject(i + 1, params[i]);
        }
                ps.executeUpdate();
        // 批量新增 Items
        String sql1 = "INSERT INTO items VALUES(NULL,?,?,?,?,?,?)";
        ps = conn.prepareStatement(sql1);
        for(Item item : itemList){
        item.setOid(oid);
        Object[] params1 = {item.getOid(),item.getBid(),item.getCreateDate(),
        item.getCount(),item.getPrice(),item.getState()};
        for (int i = 0; i < params1.length; i++) {
                ps.setObject(i + 1, params1[i]);
        }
        ps.addBatch();
        }
        ps.executeBatch();
        //book 表里减去购买书籍的库存数
        String sql3 = "UPDATE books SET stock = stock - ? WHERE bid = ?";
        ps = conn.prepareStatement(sql3);
        for(Item item : itemList){
        ps.setInt(1,item.getCount());
        ps.setInt(2,item.getBid());
        ps.addBatch();
        }
        ps.executeBatch();
        // 执行事务
        conn.commit();
```

```
        conn.setAutoCommit(true);
        closeResource();
        }catch (SQLException e){
        try {
        e.printStackTrace();
        // 事务执行过程中出现 SQL 异常则回滚数据
        conn.rollback();
        return -1;
        }catch (SQLException e1){
        e1.printStackTrace();
        return -1;
        }
        }finally {
        closeResource();
        }
        return 1;
        }
    }
```

（2）下单功能 Controller 层

下单功能包含 4 个 Controller 控件，具体内容如下。

1）ShowCartServlet：显示购物车中内容。

2）ModifyCartServlet：修改购物车中书籍购买数量，移除购物车中书籍。

3）AddItemServlet：结算下单。

4）MyOrdersServlet：显示用户所有订单。

在 Servlet 包中创建 ShowCartServlet.java 类。在该类中获取请求对象 Session 中书籍的 bid，实例化 BookSerivce 对象，通过 bid 使用 BookSerivce 中的方法查询书籍的相关数据，并将数据传到购物车页面 shopping.jsp 中显示出来，具体如示例代码 7-11 所示。

示例代码 7-11：ShowCartServlet.java

```
package com.xt.servlet;
import com.xt.Service.BookService;
import com.xt.entity.Book;
import javax.servlet.ServletException;
import javax.servlet.http.HttpServlet;
import javax.servlet.http.HttpServletRequest;
import javax.servlet.http.HttpServletResponse;
import java.io.IOException;
import java.util.ArrayList;
```

```
import java.util.HashMap;
import java.util.List;
/**
 * 显示购物车中书籍
 */
public class ShowCartServlet extends HttpServlet {
    private BookService bookSerivce = null;
    @Override
    public void init() throws ServletException {
        bookSerivce = new BookService();
    }
    @Override
    protected void doGet(HttpServletRequest req, HttpServletResponse resp) throws Servle-
tException, IOException {
        // 获取 Session 中购物车对象
        HashMap shoppingCart = (HashMap) req.getSession().getAttribute("shoppingCart");
        List<Book> bookList = new ArrayList<>();
        // 通过购物车对象中的书籍 bid 查询书籍详情
        if(shoppingCart!=null){
            for(Object key : shoppingCart.keySet()){
                Book book = bookSerivce.findBookByBID(Integer.parseInt((String) key));
                book.setCount((int)shoppingCart.get(key));
                bookList.add(book);
            }
        }
        // 在 shopping.jsp 中动态显示购物车中书籍信息
        req.getSession().setAttribute("bookList",bookList);
        req.getRequestDispatcher("shopping.jsp").forward(req,resp);
    }
}
```

在 Servlet 包中创建 ModifyCartServlet.java 类。该类接收包含操作类型 action 参数、书籍 bid 参数和书籍购买数量 count 参数的 URL。根据 action 参数判断是修改数量操作还是移除书籍操作，以分别调用不同方法操作购物车对象中该书籍的数据，具体如示例代码 7-12 所示。

示例代码 7-12：ModifyCartServlet.java

```
package com.xt.servlet;
import java.io.IOException;
```

```
import java.util.HashMap;
import javax.servlet.ServletException;
import javax.servlet.http.HttpServlet;
import javax.servlet.http.HttpServletRequest;
import javax.servlet.http.HttpServletResponse;
/**
 * 修改或移除购物车中的图书
 */
public class ModifyCartServlet extends HttpServlet {
    @Override
    protected void service(HttpServletRequest req, HttpServletResponse resp)

    throws ServletException, IOException {
      // 获取操作种类

    String action = req.getParameter("action");
      // 获取购物车对象

    HashMap shoppingCart = (HashMap)req.getSession().getAttribute("shoppingCart");
// 移除操作则删除购物车中相应 bid 键值对,修改操作则修改键值对中的 value 值
    if(action.equals("remove")){
    shoppingCart.remove(req.getParameter("bid"));
    req.getSession().setAttribute("CartBookNum", shoppingCart.size() );
    } else if(action.equals("modify")){
    shoppingCart.put(req.getParameter("bid"),Integer.parseInt(req.getParameter("count")));
    }
    req.getRequestDispatcher("showCart").forward(req, resp);
    }
}
```

在 Servlet 包中创建 AddItemServlet.java 类。该类根据请求中用户 uid 和订单总价格创建订单对象 order,根据请求中的书籍信息参数创建订单商品集合对象 itemList,实例化 OrderSerivce 对象,执行下单操作,下单成功后移除 Session 中购物车相关参数并跳转到用户订单显示页面,具体如示例代码 7-13 所示。

示例代码 7-13：AddItemServlet.java

```
package com.xt.servlet;
import java.io.IOException;
import java.math.BigDecimal;
import java.text.SimpleDateFormat;
import java.util.ArrayList;
import java.util.Date;
import java.util.List;
import javax.servlet.ServletException;
import javax.servlet.http.HttpServlet;
import javax.servlet.http.HttpServletRequest;
import javax.servlet.http.HttpServletResponse;
import com.xt.Service.OrderService;
import com.xt.entity.Item;
import com.xt.entity.Order;
/**
 * 创建订单与订单项
 */
public class AddItemServlet extends HttpServlet {
    private OrderSerivce orderSerivce=null;
    @Override
    public void init() throws ServletException {
orderSerivce = new OrderSerivce();
    }
    @Override
    protected void doPost(HttpServletRequest req, HttpServletResponse resp)
    throws ServletException, IOException {
    // 获取当前时间
    SimpleDateFormat sdf = new SimpleDateFormat("yyyy-MM-dd HH:mm:ss");
    String date = sdf.format(new Date());
     // 获取用户 uid
int uid = (int)req.getSession().getAttribute("uid");
     // 生成订单项
    Order order = new Order();
    order.setUid(uid);
    order.setCreateDate(date);
    order.setTotal_price(new BigDecimal(req.getParameter("hidden_total_price")));
    // 生成订单商品信息
```

```
    Integer bookCount = Integer.valueOf(req.getParameter("i"));
    List<Item> itemList = new ArrayList<Item>();
    for(int i = 1; i < bookCount; i++){
        // 获取订单书籍信息并保存到集合中
    String bid_str = req.getParameter("bid_" + i);
    String book_count = req.getParameter("nums_" + i);
BigDecimal book_price = new BigDecimal(req.getParameter("price_" + i));
    Item item = new Item();
    item.setBid(Integer.valueOf(bid_str));
    item.setCount(Integer.valueOf(book_count));
            item.setPrice(book_price);
            item.setState(0);
    item.setCreateDate(date);
    itemList.add(item);
    }
     // 执行下单 SQL 操作
    orderSerivce.placeOrder(itemList,order);
     // 移除 session 中的购物车对象以及书籍数目
    req.getSession().removeAttribute("shoppingCart");
    req.getSession().removeAttribute("CartBookNum");
    resp.sendRedirect("showOrder");
    }
}
```

在 Servlet 包中创建 MyOrdersServlet.java 类，在该类中获取请求对象 Session 中的用户 uid，通过 uid 查询用户所有订单并显示在 orderlist.jsp 中，具体如示例代码 7-14 所示。

示例代码 7-14：MyOrdersServlet.java

```
package com.xt.servlet;
import java.io.IOException;
import java.util.List;
import javax.servlet.ServletException;
import javax.servlet.http.HttpServlet;
import javax.servlet.http.HttpServletRequest;
import javax.servlet.http.HttpServletResponse;
import com.xt.Service.OrderService;
/**
 * 显示用户订单
 */
```

```
public class MyOrdersServlet extends HttpServlet {
    private OrderService orderSerivce = null;
    @Override
    public void init() throws ServletException {
    orderSerivce = new OrderService();
    }
    @Override
    protected void service(HttpServletRequest req, HttpServletResponse resp)
    throws ServletException, IOException {
    // 获取 session 对象中的用户 uid
    int uid = (int)req.getSession().getAttribute("uid");
    // 通过 uid 查询所有订单并传到 orderlist.jsp 中
    List orders = orderSerivce.findItemsByUID(uid,0,0);
    req.setAttribute("orders", orders);
    req.getRequestDispatcher("orderlist.jsp").forward(req, resp);
    }
}
```

编写 web.xml 文件中的 Servlet 映射关系，具体如示例代码 7-15 所示。

示例代码 7-15：Servlet 映射配置

```
<servlet>
  <description> 浏览购物车 </description>
  <servlet-name>showCart</servlet-name>
  <servlet-class>com.xt.servlet.ShowCartServlet</servlet-class>
</servlet>
<servlet-mapping>
  <servlet-name>showCart</servlet-name>
  <url-pattern>/showCart</url-pattern>
</servlet-mapping>
<servlet>
  <description> 修改购物车 </description>
  <servlet-name>ModifyCart</servlet-name>
  <servlet-class>com.xt.servlet.ModifyCartServlet</servlet-class></servlet>
<servlet-mapping>
 <servlet-name>ModifyCart</servlet-name>
  <url-pattern>/modify</url-pattern>
</servlet-mapping>
<servlet>
```

```
    <description> 生成订单 </description>
    <servlet-name>addNewOrder</servlet-name>
    <servlet-class>com.xt.servlet.AddItemServlet</servlet-class>
  </servlet>
  <servlet-mapping>
    <servlet-name> addNewOrder </servlet-name>
    <url-pattern>/newOrder</url-pattern>
  </servlet-mapping>
  <servlet>
    <description> 查看订单 </description>
    <servlet-name>showOrder</servlet-name>
    <servlet-class>com.xt.servlet.MyOrdersServlet</servlet-class>
  </servlet>
  <servlet-mapping>
    <servlet-name>showOrder</servlet-name>
    <url-pattern>/showOrder</url-pattern>
  </servlet-mapping>
```

（3）下单模块 View 层

下单功能 View 层有两个相关页面分别是 shopping.jsp 和 orderlist.jsp，其中 shopping.jsp 页面中的相关交互如下。

1）修改或删除购物车中的书籍：在 JavaScript 脚本中编写 modify 和 remove 方法，将待操作书籍的 bid 和修改后数量以参数的形式写入 ModifyCartServlet 类映射的 URL 地址中。方法被调用时，通过 URL 定向到 Controller 层的相关 ModifyCartServlet 类，并执行其中的 service 方法。

2）结算购物车：通过页面中的 form 表单将购物车中书籍信息以 Post 方法打包发送到 Web 服务器，以表单提交的 URL 定位到 AddItemServlet 类处理该数据，并将下单流程执行完毕。

编写 shopping.jsp 页面，在该页面中可以查询购物车中内容，并执行修改购买数量、移除购买书籍以及结算下单操作，具体如示例代码 7-16 所示。

示例代码 7-16：shopping.jsp

```
<%@ page language="java" import="java.util.*" contentType="text/html; charset=utf-8"
isELIgnored="false" %>
<%@ taglib uri="http://java.sun.com/jstl/core_rt" prefix="c" %>
<link rel="stylesheet" href="css/bootstrap3.css">
<link rel="stylesheet" href="css/style4.css">
<link rel="stylesheet" href="css/style.default.css" id="theme-stylesheet">
<link rel="stylesheet" href="css/1font-awesome.min.css">
<link rel="stylesheet" href="css/custom.css">
```

```
<jsp:include page="elements/index_head.jsp"/>
<script type="text/javascript">
// 修改购物车中书籍数量
   function modify(bid, i) {
      var num = document.getElementById("nums" + i);
      window.location = "modify?action=modify&bid=" + bid + "&count=" + num.value;
}
// 从购物车中移除书籍
   function remove(bid_str) {
      window.location = "modify?action=remove&bid=" + bid_str;
   }
   // 计算购物车内价格信息
   function countNum(size_str) {
      var size = parseFloat(size_str);
      var total_price = 0;
      for (var i = 1; i <= size; i++) {
         var num_str = document.getElementById("nums_" + i);
         var book_price_str = document.getElementById("price_" + i);
         var num = parseInt(num_str.value);
         var book_price = parseFloat(book_price_str.value);
         console.log(num);
         console.log(book_price)
         total_price = total_price + num * book_price;
      }
      var hidden_total_price = document.getElementById("hidden_total_price");
      hidden_total_price.value = total_price;
      document.getElementById("total_price").innerHTML = total_price;
   }
</script>
<body onload="countNum(${CartBookNum})">
<jsp:include page="elements/main_menu.jsp"/>
<div id="all">
<div id="content">
<div class="container">
<div id="basket">
<form class="box" method="post" name="shoping" action="newOrder">
<div class="table-responsive">
 <table class="table">
```

```
<thead>
  <tr>
   <th class="checker"></th>
   <th class="view"> 图片预览 </th>
   <th class="name"> 书名 </th>
   <th class="nums"> 数量 </th>
   <th class="price"> 价格 </th>
   <th class="nums"> 操作 </th>
  </tr>
  </thead>
  <tbody>
  <c:set value="1" var="i"/>
  <c:forEach var="book" items="${bookList}">
  <tr id="book_"${i}>
    <td><input type="checkbox" name="bookId" id="bookId" value="${book.bid}"/></
td>
   <input type="hidden" id="bid_${i}" name="bid_${i}" value="${book.bid}"/>
    <td class="thumb"><img src="${book.image}" width="100px" height="100px"/></
td>
   <td class="title">"${book.bookname}"</td>
   <td><input class="input-text" id="nums_${i}" name="nums_${i}" type="text"
  value="${book.count}" onchange="modify(${book.bid}, ${i})"/></td>
     <td> ￥<input id="price_${i}" name="price_${i}" type="hidden" value="${book.
price}">${book.price}</td>
   <td><span id="remove_${i}"><a class="delete" style="" href="#"onClick="remove(
  ${book.bid})"> 从购物车移除 </a></span></td>
  </tr>
  <c:set value="${i+1}" var="i"/>
  </c:forEach>
  </tbody>
  </table>
 <input type="hidden" id="i" name="i" value="${i}"/>
 <div class="button" style="float: right">
  <h4> 总价：￥<span id="total_price"></span> 元 </h4>
  <input type="hidden" id="hidden_total_price" name="hidden_total_price"/>
 </div>
 </div>
 <div class="box-footer d-flex justify-content-between flex-column flex-lg-row">
```

```
        <div class="right">
          <button class="input-chart ok" type="submit" name="submit"> 立即结算 </button>
        </div>
      </div>
    </form>
    </div>
    </div>
    </div>
    </div>
    <jsp:include page="elements/main_bottom.html"/>
```

编写 orderlist.jsp 页面，在该页面中通过 EL 表达式获取并显示用户的订单信息，具体如示例代码 7-17 所示。

示例代码 7-17：orderlist.jsp

```
    <%@ page language="java" import="java.util.*" contentType="text/html; charset=utf-8"
isELIgnored="false"%>
    <%@ taglib uri="http://java.sun.com/jstl/core_rt" prefix="c" %>
    <link rel="stylesheet" href="css/style4.css">
    <link rel="stylesheet" href="css/style3.css">
    <jsp:include page="elements/index_head.jsp"/>
    <body>
    <jsp:include page="elements/main_menu.jsp"/>
    <div id="fh5co-pricing1" class="fh5co-bg-section">
      <div class="container">
        <div class="row animate-box">
          <div class="col-md-12  text-center fh5co-heading">
            <h2> 订单详情 </h2>
          </div>
        </div>
        <div class="row">

          <div id="basket1" class="col-lg-12">
            <div class="box list orderList">
              <table class="table table-striped">
                <tr class="title">
                  <th class="orderId"> 订单编号 </th>
                  <th class="userName"> 收货人 </th>
                  <th class="createTime"> 下单时间 </th>
```

```
                <th class="status"> 总价 </th>
            </tr>
            <c:set var="oid_count" value="0" />
            <c:set var="td_id" value="0" />
            <c:forEach var="order" items="${orders}">
                <tr>
                    <c:if test="${oid_temp != order.oid}">
                        <c:set var="td_id" value="${td_id+1}" />
                        <td id="id_${td_id}">${order.oid}</td>
                        <td id="user_${td_id}">${order.username}</td>
                        <td id="crdt_${td_id}">${order.createdate}</td>
                        <td id="total_${td_id}">${order.total_price}</td>
                    </c:if>
                </tr>
            </c:forEach>
        </table>
      </div>
    </div>
  </div>
 </div>
</div>
</body>
<jsp:include page="elements/main_bottom.html"/>
```

完成上述操作后，网上书城项目下单模块功能已经完成，效果如图 7-6 和图 7-7 所示。

本次任务以 MVC 模式为架构，运用 JSP、JavaBean、JDBC、Servlet 有关知识，实现网上书城项目中书籍显示功能、搜索功能和下单功能，通过本次整合练习，掌握 JavaWeb 的体系知识及网站实现的基本技能。

model	模型	view	视图
controller	控制者		

1. 选择题

关于 MVC 架构的缺点，下列叙述不正确的是（　　）。

A. 提高了对开发人员的要求　　B. 代码复用率低

C. 增加了文件管理的难度　　D. 产生较多的文件

2. 简答题

在 MVC 模型中 M、V、C 分别指什么？简述它们之间的关系。

第八章　Vue 技术重构网上书城项目

通过学习 Vue 相关知识，结合 JSP 有关知识，掌握 Vue 技术重构 JavaWeb 项目的方法，具有运用所学习的相关知识重构网上书城项目的能力。在任务实现过程中：

- 了解 Vue 基本概念；
- 掌握 Vue 的环境搭建；
- 掌握 Vue 路由的使用方法；
- 掌握 Vue 与 Web 程序后端交互。

【情境导入】

在 JavaWeb 项目中，每次请求 JSP 页面都会加载页面相关的 js 文件、css 文件、图片文件等资源，有些请求只针对局部内容修改，却需要重新下载整个页面的静态资源，这种加载模式在一些复杂页面中会使页面功能速度变慢，从而严重影响使用的体验。

【功能描述】

- 重构网上书城项目首页。

技能点一　Vue 介绍

Vue 是一种用于搭建 Web 应用前端页面的渐进式框架。与其他前端框架不同的是，Vue 是一种自下向上逐层应用的结构，只关注 View 层，是一种既容易上手编写应用，又便于与第三方库或现有项目整合的框架。而且，Vue 与现代化的工具链以及各种支持类库结合使用时，也完全能够为复杂的单页应用提供驱动。

Vue 是一种 MVVM（Model-View-ViewModel）框架，Model 代表数据模型，定义数据操作的业务逻辑，View 代表视图层，负责将数据模型渲染到页面上，ViewModel 通过双向绑定把 View 和 Model 进行同步交互，不需要手动操作 DOM。

技能点二　构建 Vue 项目

1. 初始化 Vue 项目

Vue 项目是在 Node.js 环境下开发的，使用 Vue-CLI 作为开发系统，webpack 作为静态模块打包工具，具体步骤如下。

第一步：安装 Node.js 环境。

在 Node.js 官网下载安装包进行安装，下载地址为 https://nodejs.org/en/，如图 8-1 所示。

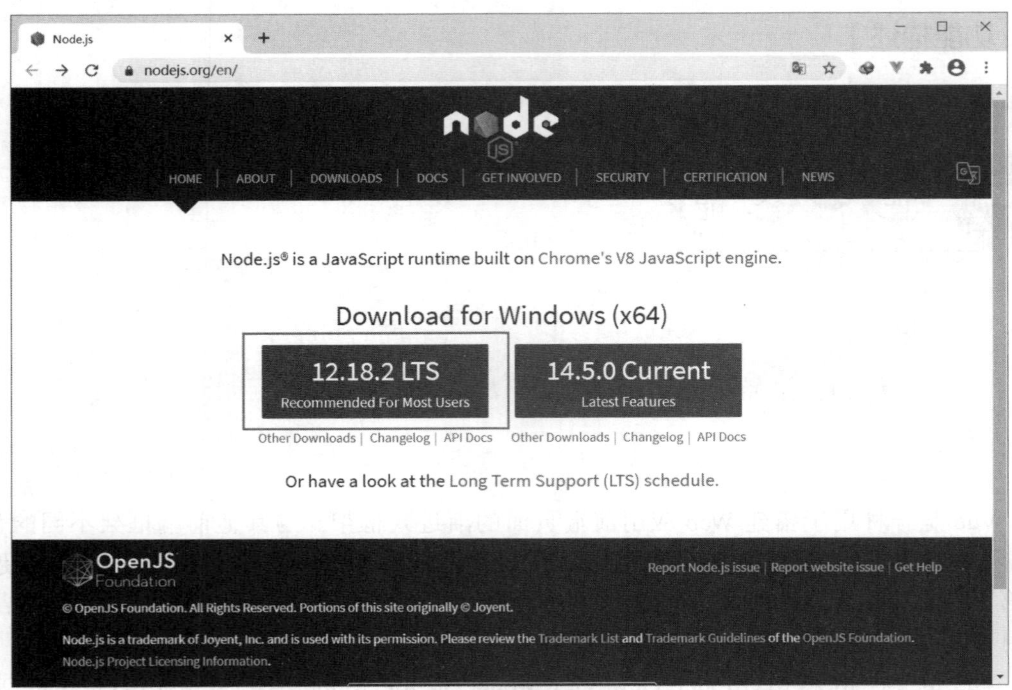

图 8-1 下载 Node.js

第二步：安装 Vue-CLI 与 webpack。

使用 npm 安装全局 Vue-CLI，安装命令如下：

```
npm install --global vue-cli
```

使用 npm 安装全局 webpack，安装命令如下：

```
npm install -g webpack
```

第三步：初始化 Vue 项目。

新建项目文件夹，使用 cmd 切换到项目文件目录下，执行 Vue 项目的初始化命令，命令代码如下：

```
vue init webpack
```

项目初始化过程选项如图 8-2 所示。

```
npm
F:\vue\JavaBook>vue init webpack javabook

 Project name javabook
 Project description A Vue.js project
 Author zt
 Vue build standalone
 Install vue-router? Yes
 Use ESLint to lint your code? No
 Set up unit tests No
 Setup e2e tests with Nightwatch? No
 Should we run `npm install` for you after the project has been created? (recommended) npm

  vue-cli · Generated "javabook".
```

图 8-2 项目初始化过程选项

选项详解如下。

1）Project name：项目名称，不能有大写字母。

2）Project description：项目描述，根据个人需要填写。

3）Author：作者名称，默认为计算机用户名。

4）Vue build：项目构建方式，选择 Runtime + Compiler（运行时 + 编译器）选项。另一种 Runtime-only（只包含运行时）方式构建速度更快，但不能识别非 Vue 文件中的 template。

5）Install vue-router：是否安装 Vue 的路由插件。

6）Use ESLint to lint your code：是否使用 ESLint 检测代码，ESLint 是一个语法规则和代码风格的检查工具，可以用来保证写出语法正确、风格统一的代码，一般在多人项目中使用。

7）Set up unit tests：是否安装单元测试。

8）Setup e2e tests with Nightwatch：是否安装 E2E 测试框架 NightWatch。

9）Should we run 'npm install' for you after the project has been created：选择使用 npm 安装、yarn 安装或手动安装。

2.IDEA 开发 Vue

在 IDEA 中安装 Vue 项目 Plugins 插件 Vue.js 和 Node.js，安装过程如图 8-3 和图 8-4 所示。

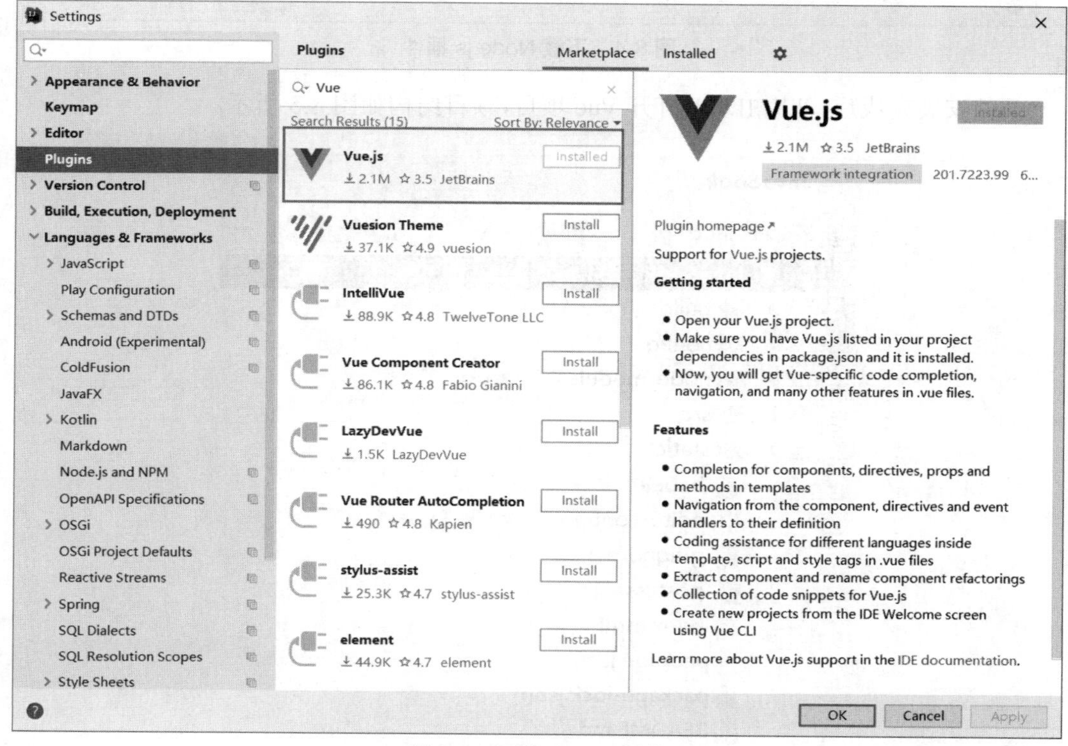

图 8-3　下载 Vue.js 插件

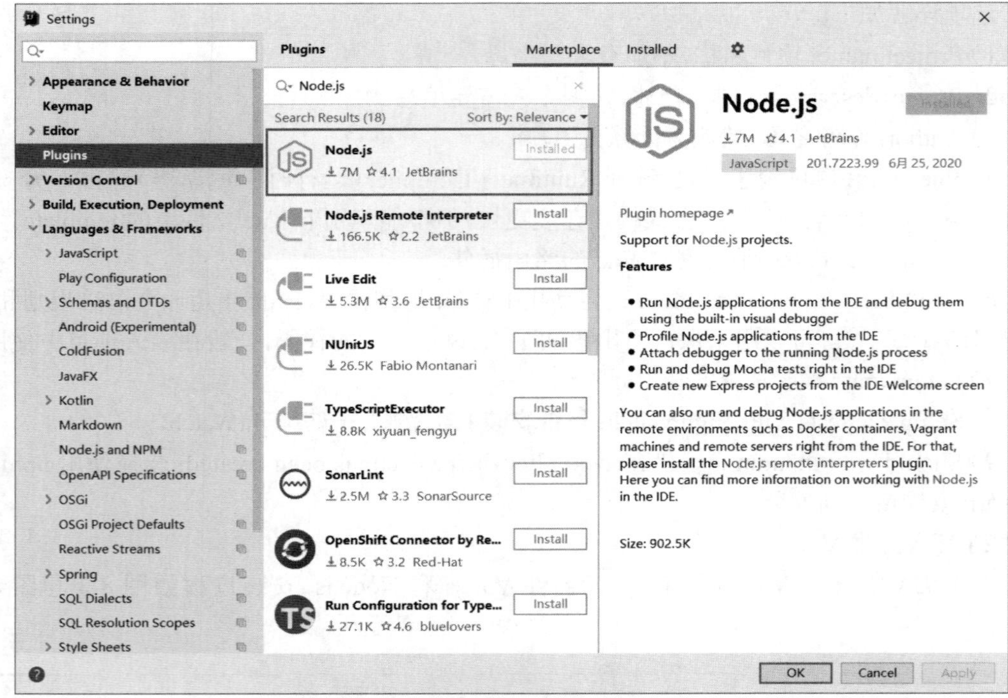

图 8-4　下载 Node.js 插件

插件安装完成后，使用 IDEA 打开 Vue 项目，项目结构如图 8-5 所示。

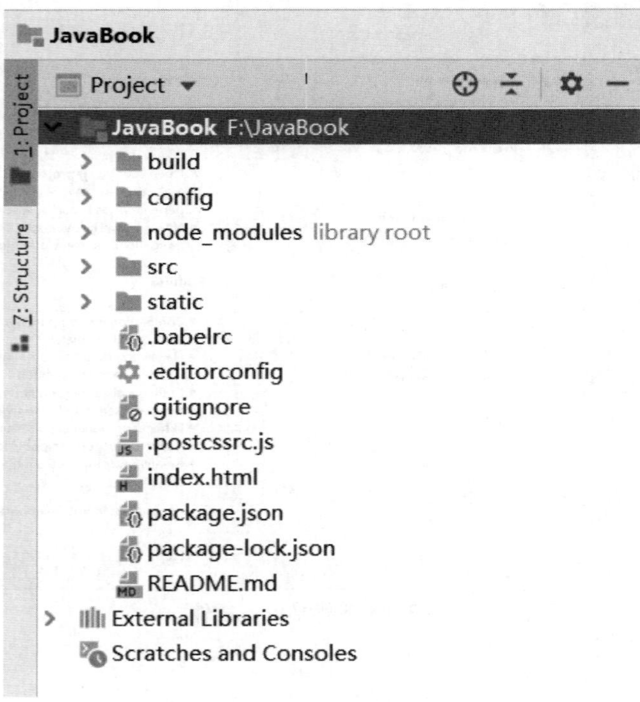

图 8-5　项目结构

项目的文件结构如表 8-1 所示。

表 8-1　项目文件结构

名称	作用
build	存放编译时用到的脚本文件
config	存放 Vue 项目的各种配置
node_modules	存放 node 第三方包
src	存放项目源代码
static	存放项目静态文件
package.json	项目配置文件

配置运行命令，配置过程如图 8-6 和图 8-7 所示。

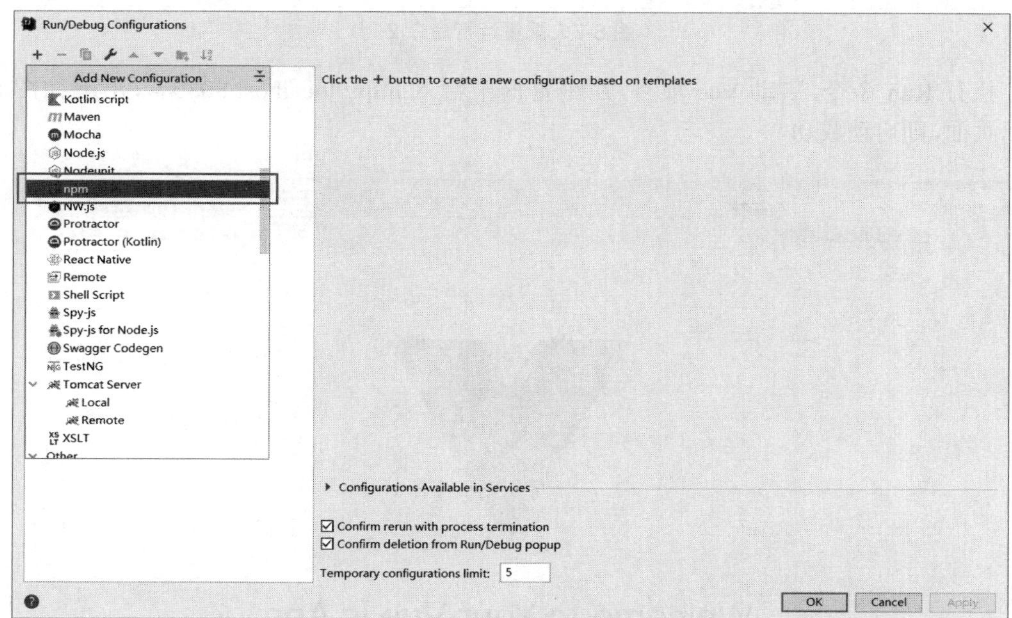

图 8-6　配置运行命令 1

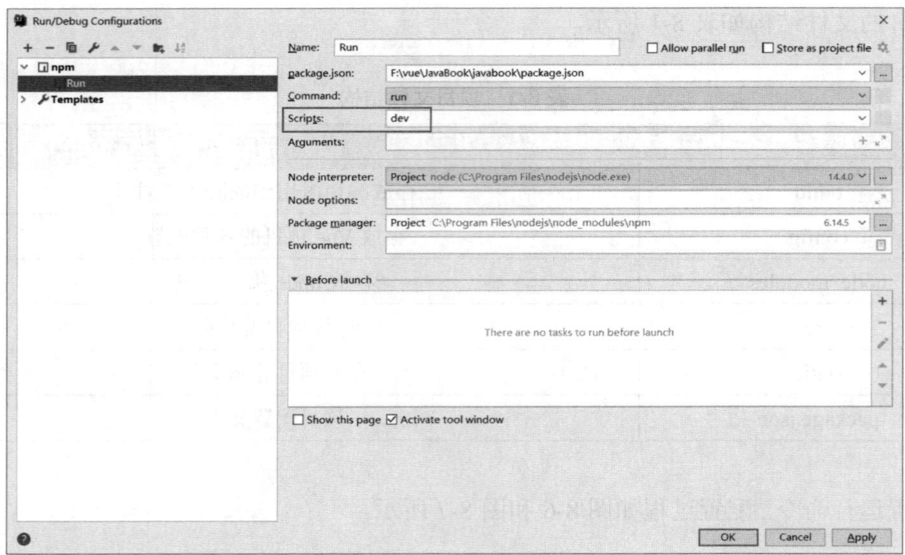

图 8-7 配置运行命令 2

执行 Run 命令，启动 Vue 项目，在浏览器中输入 http://localhost:8080/#/，出现如图 8-8 所示页面，则启动成功。

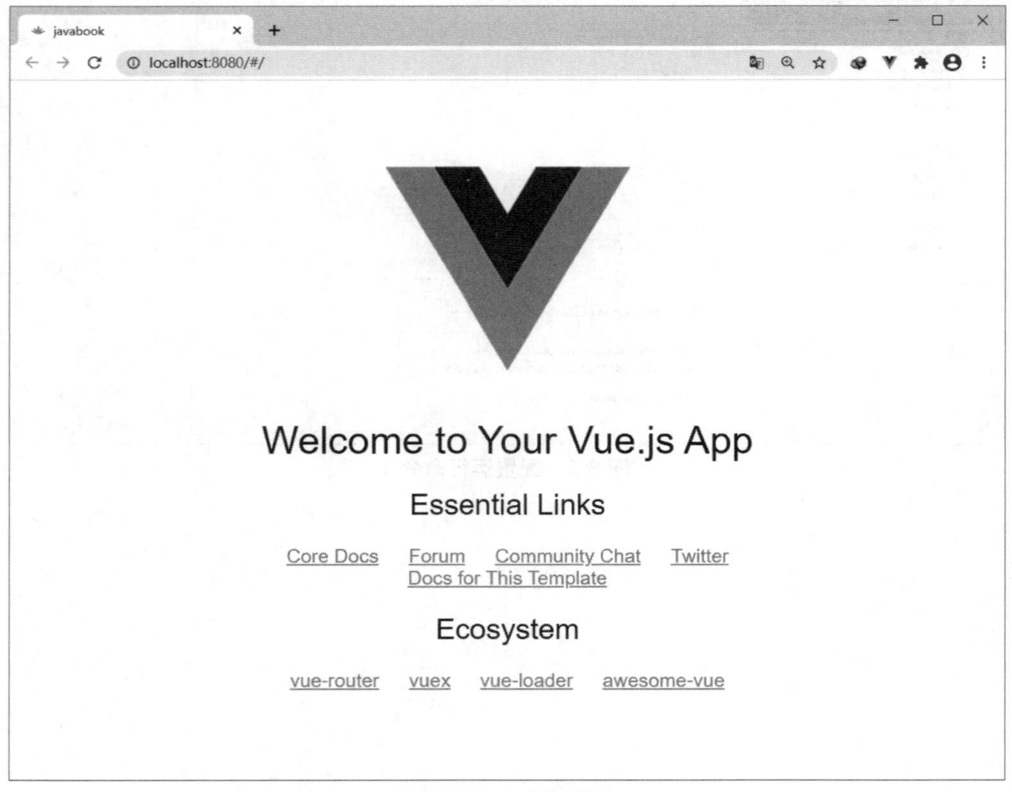

图 8-8 Vue 项目首页

技能点三　Vue 路由

Vue 路由是一种前端路由，即浏览器 URL 地址变化后，并不会发送相关请求到后端服务器，而是由 JavaScript 监听响应的请求，并更新视图页面。

Vue 路由在 src/router/index.js 中进行配置，在该 js 文件中使用 import 引入组件，代码如下。

```
import Demo from '@/components/ComponentDemo'
```

文件中实例化了路由对象 Router，在该对象中可对路由进行相关配置，代码如下。

```
export default new Router({
 routes: [
  {
   path:'/demo',    // 配置路由地址
   name:'Demo'   // 设置路由名称
   components: {
    Demo, // 配置路由中的组件
   }
  }
 ]
})
```

路由配置完成后，在 Vue 程序的主组件 Vue.app 中，使用 <router-view> 标签标记路由匹配到组件的渲染位置，<router-view> 标签的使用方法如下。

```
<template>
 <div id="app">
  <router-view name="Demo"/>  //name 属性为此处渲染的组件名
 </div>
</template>
```

【实例】使用路由将 2 个组件拼接成一个完整网页，网页显示效果如图 8-9 所示。

图 8-9　路由实例

第一步：编写组件头部组件 Head.vue，如示例代码 8-1 所示。

示例代码 8-1：Head.vue

```
<template>
  <p> 网站头部组件 </p>
</template>
<script>
  export default {
    name: "Head"
  }
</script>
```

第二步：编写主页面组件 Main.vue，如示例代码 8-2 所示。

示例代码 8-2：Main.vue

```
<template>
  <p> 网页主页面组件 </p>
</template>
<script>
  export default {
    name: "Main"
  }
</script>
```

第三步：在 src/router/index.js 中为组件设置路由，如示例代码 8-3 所示。

示例代码 8-3：src/router/index.js

```
import Vue from 'vue'
import Router from 'vue-router'
import Head from '@/components/Head'
import Main from '@/components/Main'
Vue.use(Router)
export default new Router({
  routes: [
    {
      path:'/demo',
      name:'Demo',
      components: {
        Head,
        Main,
      }
    }
  ]
})
```

第四步：在 Vue 程序的主组件 App.vue 中设置组件渲染位置，如示例代码 8-4 所示。

示例代码 8-4：App.vue

```
<template>
  <div id="app">
    <router-view name="Head"/>
    <p>============</p>
    <router-view name="Main"/>
  </div>
</template>
<script>
export default {
  name: 'App'
}
</script>
<style>
#app {
  font-family: 'Avenir', Helvetica, Arial, sans-serif;
  -webkit-font-smoothing: antialiased;
```

```
    -moz-osx-font-smoothing: grayscale;
    text-align: center;
    color: #2c3e50;
    margin-top: 60px;
  }
</style>
```

技能点四　Vue 与 Web 程序后端交互

使用 Vue 构建前端页面的项目，后端程序不再参与渲染页面，只需要提供数据响应，页面所有渲染过程全部在前端进行。

1.JSON 数据处理

JavaWeb 程序后端与 Vue 一般使用 JSON 字符串进行数据传输，在 JavaWeb 项目中引入 fastjson 依赖包来处理 JSON 字符串和业务对象之间的转化，引入步骤如下。

第一步：通过 https://mvnrepository.com/artifact/com.alibaba/fastjson 页面下载 fastjson 库 jar 包。

第二步：将 jar 包粘贴进项目中的 WEB-INF/lib 文件夹中。

fastjson 将 Java 对象转化为 JSON 格式的常用方法如下：

```
String jsonObject = JSON.toJSONString(JavaObject)
```

【实例】使用 fastjson 包中方法将 Java 对象转化为 JSON 字符串输出，效果如图 8-10 所示。

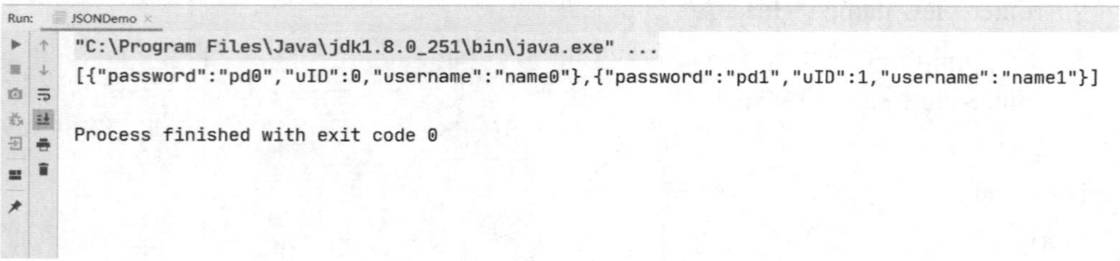

```
Run: JSONDemo
"C:\Program Files\Java\jdk1.8.0_251\bin\java.exe" ...
[{"password":"pd0","uID":0,"username":"name0"},{"password":"pd1","uID":1,"username":"name1"}]

Process finished with exit code 0
```

图 8-10　JSON 数据转换

实现代码如示例代码 8-5 所示。

示例代码 8-5：JSON 数据转换

```
package com.xt.json;
import com.alibaba.fastjson.JSON;
import com.xt.entity.UserInfo;
```

```
import java.util.ArrayList;
import java.util.List;
public class JSONDemo {
   public static void main(String[] args) {
      List<UserInfo> userList = new ArrayList<UserInfo>();
      for(int i=0;i<2;i++){
         UserInfo user = new UserInfo();
         user.setUID(i);
         user.setUsername("name"+i);
         user.setPassword("pd"+i);
         userList.add(user);
      }
      System.out.println(JSON.toJSONString(userList));
   }
}
```

2.Axios 应用

Vue 本身并没有向后端发送请求的功能，需要引入第三方支持才能实现。vue-resource 库与 axios 都可以使 Vue 项目实现 ajax 异步请求，本书主要对 axios 进行介绍。

在 Vue 项目中使用以下命令安装 axios 依赖：

```
npm install axios
```

修改 src/main.js 文件，新增以下命令启动 axios：

```
import axios from "axios"
Vue.prototype.$axios = Axios;
```

axios 执行 get 请求的方法如下：

```
//get 请求格式 1
axios.get('/book?uid=1')        //get 请求 url
 .then(function (response) {   // 处理返回的响应
  console.log(response);
 })
 .catch(function (error) {      // 处理请求的异常
  console.log(error);
 });

//get 请求格式 2
axios.get('/book', {        //get 请求 url（不含请求参数）
```

```
    params: {           // 在 get 请求中添加参数
      uid: 1
    }
  })
  .then(function (response) {     // 处理返回的响应
    console.log(response);
  })
  .catch(function (error) {       // 处理请求的异常
    console.log(error);
  });
```

执行 post 请求的方法如下：

```
axios.post('/book', {         //post 请求 url
    bookname: ' 奥德赛 ',       //post 请求参数
    price: '35,
  })
  .then(function (response) {   // 处理返回的响应
    console.log(response);
  })
  .catch(function (error) {     // 处理请求的异常
    console.log(error);
  });
```

在使用 Vue 进行项目开发的过程中，Vue 项目与 JavaWeb 后端运行在不同的端口上，Vue 向 JavaWeb 请求数据，需要在 config/index.js 文件夹下设置跨域请求，代码如下。

```
dev: {
    assetsSubDirectory: 'static',
    assetsPublicPath: '/',
    proxyTable: {   // 设置代理
      '/api':{       // 匹配项，写在 url 中即激活代理
        target:'http://localhost:8090,  // 设置代理调用的域名和端口号
        changeOrigin: true,  // 是否允许跨域
        pathRewrite:{      // 地址复写
          '^/api':''
        }
      }
    },
```

其中，pathRewrite 属性中的设置是将请求 URL 中匹配项 /api 部分替换为空，如请求 URL 为 /api/books，实际请求地址将被复写成 http://localhost:8090/books。

【实例】编写实例请求后端数据，后端发送数据如图 8-11 所示，具体操作步骤如下。

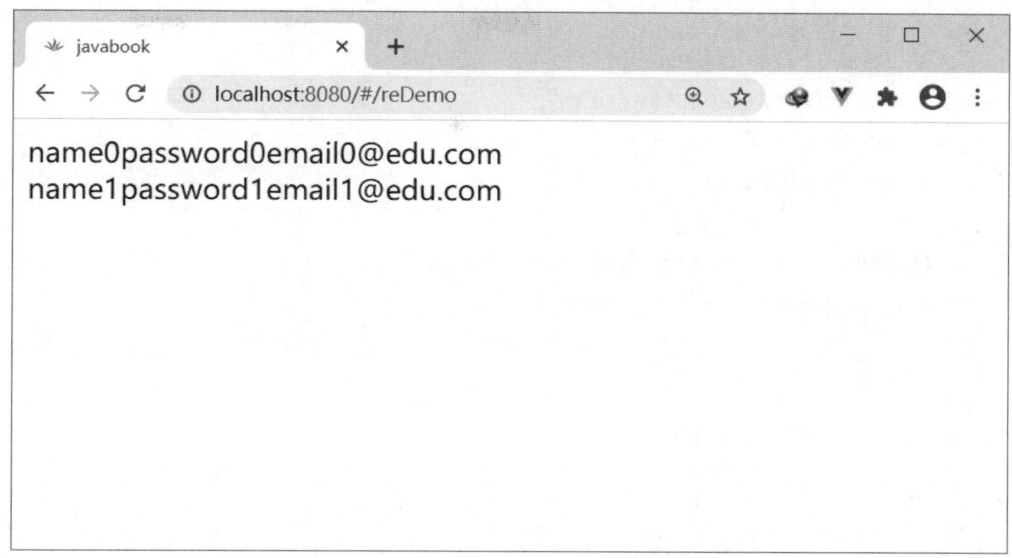

图 8-11　axios 请求实例

第一步：编写后端 Servlet，代码如示例代码 8-6 所示。

示例代码 8-6：DemoServlet.java

```
package com.xt.servlet;
import com.alibaba.fastjson.JSON;
import com.xt.entity.UserInfo;
import javax.servlet.ServletException;
import javax.servlet.http.HttpServlet;
import javax.servlet.http.HttpServletRequest;
import javax.servlet.http.HttpServletResponse;
import java.io.IOException;
import java.io.OutputStream;
import java.util.ArrayList;
import java.util.List;
public class DemoServlet extends HttpServlet {
    @Override
        protected void service(HttpServletRequest request, HttpServletResponse response)
throws ServletException, IOException {
        response.setContentType("text/json;charset=UTF-8");
        response.setCharacterEncoding("UTF-8");
      List<UserInfo> userList = new ArrayList<UserInfo>();
        for(int i=0;i<2;i++){
```

```
            UserInfo user = new UserInfo();
            user.setUID(i);
            user.setUsername("name"+i);
            user.setPassword("password"+i);
            user.setEmail("email"+i+"@edu.com");
            userList.add(user);
        }
        OutputStream out = response.getOutputStream();
        out.write(JSON.toJSONBytes(userList));
    }
}
```

该 Servlet 的映射配置如下：

```
<servlet>
  <servlet-name>DemoSer</servlet-name>
  <servlet-class>com.xt.servlet.DemoServlet</servlet-class>
</servlet>
<servlet-mapping>
  <servlet-name>DemoSer</servlet-name>
  <url-pattern>/demoSer</url-pattern>
</servlet-mapping>
```

第二步：编写请求数据组件 RequestDemo.vue，如示例代码 8-7 所示。

示例代码 8-7：RequestDemo.vue

```
<template>
  <div>
    <tr v-for="user in UserList">        // 循环遍历 UserList 对象
      <td>{{user.username}}</td>       // 输出 user 中的数据
      <td>{{user.password}}</td>
      <td>{{user.email}}</td>
    </tr>
  </div>
</template>
<script>
  export default {
    name: "RequestDemo",
    data:function(){
      return{
        UserList:[],   // 定义 UserList 数组
```

```
        }
      },
      mounted(){
      // 向后端发送请求
       this.$axios.get('/api/BookShop/demoSer').then((response)=>{ // 处理响应数据
         this.UserList = response.data;  // 将响应的 JSON 串赋值给 UserList 对象
       }).catch(function (error) {
          console.log(error);
        });
      },
    }
</script>
```

第三步：为组件设置路由，如示例代码 8-8 所示。

示例代码 8-8：路由配置文件 index.js

```
import Vue from 'vue'
import Router from 'vue-router'
import RequestDemo from '@/components/RequestDemo'
Vue.use(Router)
export default new Router({
  routes: [
    {
      path:'/reDemo',
      name:'RequestDemo',
      components: {
        RequestDemo,
      }
    }
  ]
})
```

第四步：在程序主组件中设置渲染位置，如示例代码 8-9 所示。

示例代码 8-9：配置主组件 APP.vue

```
<template>
  <div id="app">
    <router-view name="RequestDemo"/>
  </div>
</template>
<script>
```

```
export default {
  name: 'App'
}
</script>
```

3.Session 应用

Vue 本身不支持对 Session 的操作，需添加第三方依赖 vue-session 来完善这项功能，安装代码如下：

```
npm install vue-session
```

修改 src/main.js 文件，新增以下命令启动 vue-session：

```
import VueSession from "vue-session";
Vue.use(VueSession)
```

在 Vue 项目开发测试时，Session 跨域传输需使用 axios 的证书功能，在 src/main.js 文件中编写如下代码：

```
axios.defaults.withCredentials = true;
```

vue-session 常用方法如下：

```
this.$session.set(key,value)  // 新增 session
this.$session.remove(key)   // 根据 key 移除 session
```

【实例】使用 vue-session 新增、删除 Session，观察浏览器中 Session 内容变化，效果如图 8-12 和图 8-13 所示。

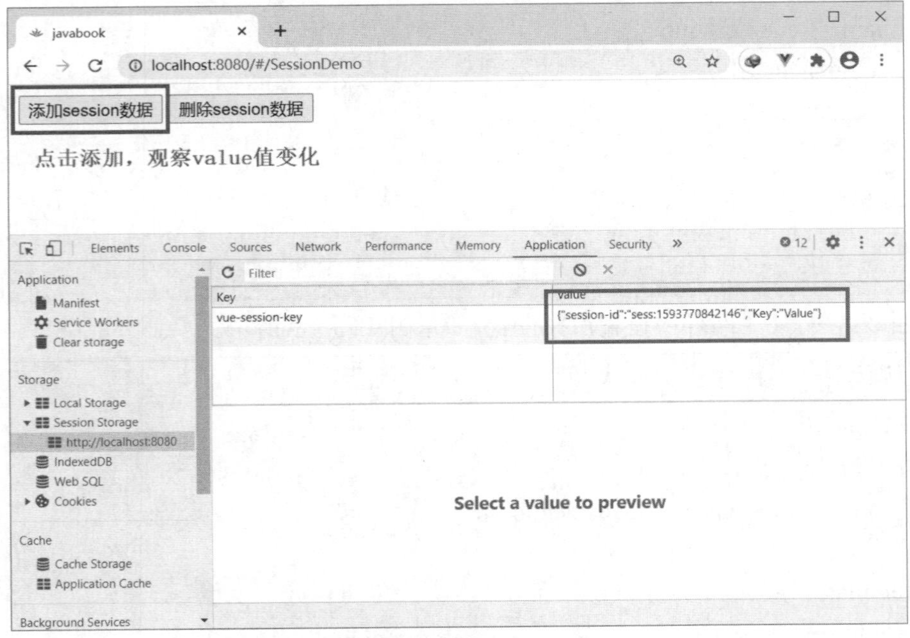

图 8-12 添加 Session

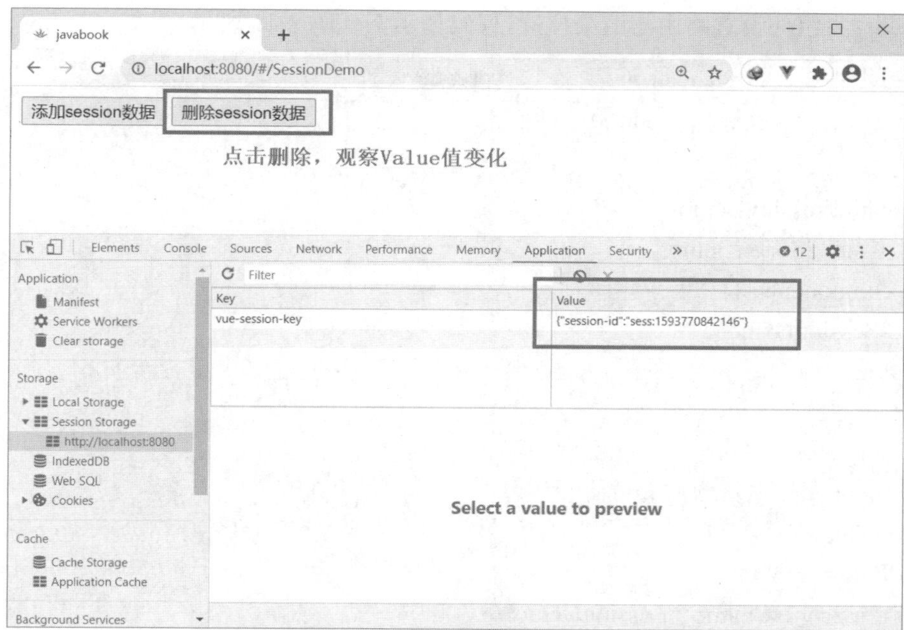

图 8-13　删除 Session

第一步：新建组件 SessionDemo.vue，代码如示例代码 8-10 所示。

示例代码 8-10：SessionDemo.vue

```
<template>
  <div>
    <input type="button" value=" 添加 session 数据 " v-on:click="sessionIN">
    <input type="button" value=" 删除 session 数据 " v-on:click="deleteSession">
  </div>
</template>
<script>
  export default {
    name: "SessionDemo",
    methods:{
      sessionIN:function () {        // 新增 session 方法
        this.$session.set("Key","Value")
      },
      deleteSession:function () {    // 删除 session 方法
        this.$session.remove("Key");
      }
    }
  }
</script>
```

第二步：为组件设置路由与渲染位置，代码如示例代码 8-11 所示。

示例代码 8-11：src/router/index.js 与 App.vue 设置

```
// 在 src/router/index.js 中设置路由
{
    path:'/sessionDemo',
    name:'SessionDemo',
    components: {
      SessionDemo,
    }
  }

// 在 App.vue 中新增渲染位置设置
<template>
  <div id="app">
    <router-view name="SessionDemo"/>
  </div>
</template>
```

运用 Vue 知识，重构网上书城项目首页。

第一步：在 src/main.js 文件中设置并启用 axios 与 vue-session 支持，代码如示例代码 8-12 所示。

示例代码 8-12：main.js

```
import Vue from 'vue'
import Axios from "axios"
import App from './App'
import router from './router'
import VueSession from "vue-session";
Vue.config.productionTip = false
Vue.prototype.$axios = Axios;
Vue.use(router)
Vue.use(VueSession)
Axios.defaults.withCredentials = true;
new Vue({
```

```
    el: '#app',
    router,
    components: { App },
    template: '<App/>',
    data () {
      return {
        message:'Hello',
        info: null
      }
    },
  })
```

第二步：创建 main_menu 菜单组件，代码如示例代码 8-13 所示。

示例代码 8-13：main_menu.vue

```
<template>
  <div class="hero-content">
    <header class="site-header">
      <div class="top-header-bar">
        <div class="container-fluid">
          <div class="row">
            <div class="col-12 col-lg-6 d-none d-md-flex flex-wrap justify-content-center justify-content-lg-start mb-3 mb-lg-0">
              <div class="header-bar-email d-flex align-items-center">
                <font color="BLACK"> 欢迎您，{{loginuser}}</font>   
              </div>
              <div class="header-bar-text lg-flex align-items-center">
                <p> 在线人数：1</p>
              </div>
            </div>
            <div class="col-12 col-lg-6 d-flex flex-wrap justify-content-center justify-content-lg-end align-items-center">
              <div class="header-bar-search">
                <form class="flex align-items-stretch" method="post" name="search" action="-search">
                  <input type="search" placeholder=" 输入您想搜索的书籍 ?" name="keywords" >
                  <button type="submit" value="" class="flex justify-content-center align-items-center"> 搜索 </button>
```

```
                </form>
              </div>
              <div class="header-bar-menu">
                <ul class="flex justify-content-center align-items-center py-2 pt-md-0">
                  <li><a onclick="window.location='register.jsp';" > 注册 </a></li>
                  <li><a onclick="window.location='logout.jsp';" > 登录 </a></li>
                </ul>
              </div>
            </div>
          </div>
        </div>

      </div>
      <div class="nav-bar">
        <div class="container">
          <div class="row">
            <div class="col-9 col-lg-3">
              <div class="site-branding">
                  <h1 class="site-title"><a href="index.html" rel="home">Line<span>Book</
span></a></h1>
              </div>
            </div>
            <div class="col-3 col-lg-9 flex justify-content-end align-content-center">
              <nav class="site-navigation flex justify-content-end align-items-center">
                <ul class="flex flex-column flex-lg-row justify-content-lg-end align-content-cen-
ter">
                  <li><a href="books"> 首页 </a></li>
                  <li><a href="showOrder?username=${loginuser}"> 我的订单 </a></li>
                  <li><a href="showCart"> 购物车 </a></li>
                </ul>
                <div class="hamburger-menu d-lg-none">
                  <span></span>
                  <span></span>
                  <span></span>
                  <span></span>
                </div>
                <div class="header-bar-cart">
                  <a href="#" class="flex justify-content-center align-items-center"></a>
```

```
            </div>
          </nav>
        </div>
      </div>
    </div>
  </div>
 </header>
</div>
</template>
<script>
  export default {
    name: "menu",
    data:function () {
      return{
        loginuser:'',
      }
    },
    mounted(){
      this.$session.set('loginuser','admin'),
      this.loginuser = this.$session.get('loginuser')
    }
  }
</script>
<style scoped>
 @import "../assets/css/bootstrap2.min.css";
 @import "../assets/css/style3.css";
 @import "../assets/css/elegant-fonts.css";
 @import "../assets/css/themify-icons.css";
 @import "../assets/css/swiper.min.css";
 @import "../assets/css/font-awesome.2min.css";
</style>
```

第三步：创建底部组件，如示例代码 8-14 所示。

示例代码 8-14：index_bottom.vue

```
<template>
 <div id="footer" class="wrap">
  2020&copy; 网上书城
 </div>
```

```
</template>
<script>
  export default {
    name: "index_bottom"
  }
</script>
<style scoped>
 @import "../assets/css/style.css";
</style>
```

第四步：创建首页书籍显示部分组件，如示例代码 8-15 所示。

示例代码 8-15：main.vue

```
<template>
 <div>
   <div id="carousel-example-generic" class="carousel slide center-block" data-ride="carousel">
     <ol class="carousel-indicators">
       <li data-target="#carousel-example-generic" data-slide-to="0" class="active"></li>
       <li data-target="#carousel-example-generic" data-slide-to="1"></li>
       <li data-target="#carousel-example-generic" data-slide-to="2"></li>
       <li data-target="#carousel-example-generic" data-slide-to="3"></li>
       <li data-target="#carousel-example-generic" data-slide-to="4"></li>
     </ol>
     <div class="carousel-inner" role="listbox">
       <div class="item active">
         <img src="@/assets/images/01.jpg" alt="...">
         <div class="carousel-caption">
         </div>
       </div>
       <div class="item">
         <img src="@/assets/images/02.jpg" alt="...">
         <div class="carousel-caption">
         </div>
       </div>
       <div class="item">
         <img src="@/assets/images/03.jpg" alt="...">
         <div class="carousel-caption">
         </div>
```

```
          </div>
          <div class="item">
            <img src="@/assets/images/04.jpg" alt="...">
            <div class="carousel-caption">
            </div>
          </div>
        </div>

          <a class="left carousel-control" href="#carousel-example-generic" role="button" da-
ta-slide="prev">
          <span class="glyphicon glyphicon-chevron-left" aria-hidden="true"></span>
          <span class="sr-only">Previous</span>
        </a>
         <a class="right carousel-control" href="#carousel-example-generic" role="button" da-
ta-slide="next">
          <span class="glyphicon glyphicon-chevron-right" aria-hidden="true"></span>
          <span class="sr-only">Next</span>
        </a>
      </div>
      <div id="fh5co-pricing" class="fh5co-bg-section">
        <div class="container">
          <div class="row animate-box">
            <div class="col-md-12  text-center fh5co-heading">
              <h2> 书籍列表 </h2>
            </div>
          </div>
          <div class="row">
            <div id="basket1" class="col-lg-12">
              <div class="box">
                <form>
                  <div class="table-responsive">
                    <table class="table">
                      <thead>
                      <tr class="title">
                        <th class="checker"></th>
                        <th> 书名 </th>
                        <th> 简介 </th>
                        <th> 编号 </th>
```

```
                <th class="price"> 价格 </th>
                <th class="store"> 库存 </th>
                <th class="view"> 图片预览 </th>
              </tr>
              </thead>
              <tbody>
              <tr v-for="book in bookList">
                <td><input type="checkbox" name="bookId" id="bookId" v-model="bids"
v-bind:value="book.bid"/></td>
                <td>{{book.bookname}}</td>
                <td> 书本简介 </td>
                <td> 书籍编号 </td>
                <td> ￥{{book.price}}</td>
                <td>{{book.stock}}</td>
                            <td class="thumb"><img :src="book.image" width="100px"
height="100px"/></td>
              </tr>
              </tbody>
            </table>
          </div>
          <div class="box-footer d-flex justify-content-between flex-column flex-lg-row">
            <div class="button" >
              <input class="input-btn add" type="button" name="submit" value=" 添加到
购物车 " v-on:click="cart"/>
            </div>
          </div>
        </form>
      </div>
    </div>
   </div>
  </div>
 </div>
</div>
</template>
<script>
  export default {
    name: "main",
    data:function () {
```

```
      return{
        bookList:[],
        bids:[],
      }
    },
    mounted(){
      this.$axios.get('/api/BookShop/books').then((response)=>{
        this.bookList = response.data;
      })
        .catch(function (error) {
          console.log(error);
        });
    },
    methods:{
      cart:function () {
        for(var i=0;i<this.bids.length;i++){
          this.$session.set(this.bids[i],1);
        }
        this.$router.push('/success')
      }
    }
  }
</script>
<style scoped>
  @import "../assets/css/style.default.css";
  @import "../assets/css/style4.css";
  @import "../assets/css/icomoon.css";
  @import "../assets/css/bootstrap.css";
  @import "../assets/css/custom.css";
</style>
```

第五步：创建添加购物车成功组件，如示例代码 8-16 所示。

示例代码 8-16：success.vue

```
<template>
  <div>
    <body></body>
    <div id="content" class="wrap">
      <div class="success">
```

```
        <div class="information">
          <p><a class="add" href="showCart"> 点此查看购物车详情 &gt;&gt;</a></p>
        </div>
      </div>
    </div>
  </div>
</template>

<script>
  export default {
    name: "success"
  }
</script>

<style scoped>
  @import "../assets/css/style.css";
  @import "../assets/css/style3.css";
</style>
```

第六步：配置 router 路由，如示例代码 8-17 所示。

示例代码 8-17：src/router/index.js

```
import Vue from 'vue'
import Router from 'vue-router'
import Bottom from '@/components/index_bottom'
import Menu from '@/components/main_menu'
import Main from '@/components/main'
import Success from '@/components/success'
Vue.use(Router)
export default new Router({
  routes: [
    {
      path: '/',
      components: {
        Bottom,
        Menu,
        Main,
      }
    },
```

```
    {
      path: '/success',
      components: {
        Bottom,
        Menu,
        Success,
      }
    },
  ]
})
```

第七步：配置主组件 App.vue，如示例代码 8-18 所示。

示例代码 8-18:App.vue

```
<template>
  <div id="app">
    <router-view name="Menu"/>
    <router-view name="Main"/>
    <router-view name="Success"/>
    <router-view name="Bottom"/>
  </div>
</template>
<script>
export default {
  name: 'App'
}
</script>

<style>
#app {
  background: #f0f0f09c;
}
</style>
```

第八步：修改显示书籍 servlet 类 ShowBooksServlet.java，如示例代码 8-19 所示。

示例代码 8-19: ShowBooksServlet.java

```
package com.xt.Servlet;
import java.io.IOException;
import java.io.PrintWriter;
```

```
import java.util.List;
import javax.servlet.ServletException;
import javax.servlet.http.HttpServlet;
import javax.servlet.http.HttpServletRequest;
import javax.servlet.http.HttpServletResponse;
import com.alibaba.fastjson.JSON;
import com.alibaba.fastjson.JSONObject;
import com.xt.Service.BookService;
import com.xt.util.PageTools;
/*
 * 显示全部图书
 */
public class ShowBooksServlet extends HttpServlet {
    private BookService bookSerivce = null;
    @Override
    public void init() throws ServletException {
    bookSerivce = new BookService();
    }
    @Override
    protected void service(HttpServletRequest req, HttpServletResponse resp)
    throws ServletException, IOException {
    resp.setContentType("text/json;charset=UTF-8");
    resp.setCharacterEncoding("UTF-8");
    String NO_str = req.getParameter("current_books_NO");
    int NO = NO_str==null?1:Integer.valueOf(NO_str);
    List books =  bookSerivce.findBooks(PageTools.book_num, NO);
    JSONObject jsonObject = new JSONObject();
    jsonObject.put("books",books);
    PrintWriter out = resp.getWriter();
    out.println(JSON.toJSONString(books));
    }
}
```

完成上述操作后，网上书城项目首页重构已经完成，效果如图 8-14 和图 8-15 所示。

图 8-14 Vue 重构后的主页面

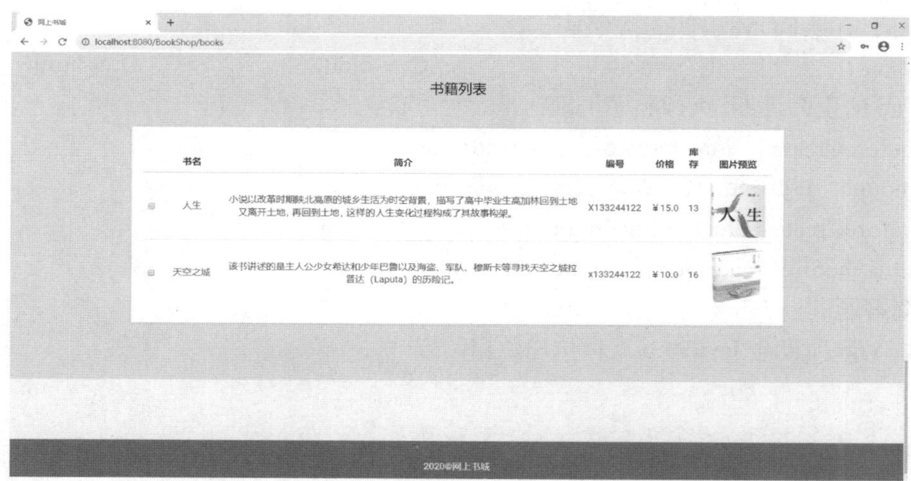

图 8-15 Vue 重构后的书籍列表

任务总结

本次任务通过对 Vue 基本知识的学习，重点熟悉 Vue 项目的构建和路由，学习 Vue 与 Web 程序后端交互的方法，了解 JSON 数据处理、Axios 应用和 Session 应用，为重构 JavaWeb 项目打下良好的基础。

英语角

template	模板	proxy	代理
router	路由器	node	节点

1. 选择题

1）以下不是 vue-router 导航钩子的是（　　）。

A. 全局导航钩子　B. 组件内的钩子　C. 页面钩子　D. 单独路由独享组件

2）以下获取动态路由 { path: '/user/:id' } 中 id 值正确的是（　　）。

A. this.$route.params.id　B. this.route.params.id

C. this.$router.params.id　D. this.router.params.id

3）以下选项中不可以进行路由跳转的是（　　）。

A. push()　B. replace()　C. route-link()　D. jump()

4）用于监听 DOM 事件的指令是（　　）。

A. v-on　B. v-model　C. v-bind　D. v-html

5）vue 的生命周期，执行顺序正确的是（　　）。

A. beforeCreate -> init->create->mount->destory

B. mount-> init->beforeCreate->create->destory

C. beforeCreate->create->init->mount->destory

D. init->beforeCreate->create->init->destroy

2. 简答题

简述 Vue 项目与 JavaWeb 项目传递数据的方式。